Y.-P. Tan, K. H. Yap, L. Wang (Eds.)

Intelligent Multimedia Processing with Soft Computing

Studies in Fuzziness and Soft Computing, Volume 168

Editor-in-chief
Prof. Janusz Kacprzyk
Systems Research Institute
Polish Academy of Sciences
ul. Newelska 6
01-447 Warsaw
Poland
E-mail: kacprzyk@ibspan.waw.pl

Further volumes of this series can be found on our homepage: springeronline.com

Vol. 152. J. Rajapakse, L. Wang (Eds.)
Neural Information Processing: Research and Development, 2004
ISBN 3-540-21123-3

Vol. 153. J. Fulcher, L.C. Jain (Eds.)
Applied Intelligent Systems, 2004
ISBN 3-540-21153-5

Vol. 154. B. Liu
Uncertainty Theory, 2004
ISBN 3-540-21333-3

Vol. 155. G. Resconi, J.L. Jain
Intelligent Agents, 2004
ISBN 3-540-22003-8

Vol. 156. R. Tadeusiewicz, M.R. Ogiela
Medical Image Understanding Technology, 2004
ISBN 3-540-21985-4

Vol. 157. R.A. Aliev, F. Fazlollahi, R.R. Aliev
Soft Computing and its Applications in Business and Economics, 2004
ISBN 3-540-22138-7

Vol. 158. K.K. Dompere
Cost-Benefit Analysis and the Theory of Fuzzy Decisions – Identification and Measurement Theory, 2004
ISBN 3-540-22154-9

Vol. 159. E. Damiani, L.C. Jain, M. Madravia
Soft Computing in Software Engineering, 2004
ISBN 3-540-22030-5

Vol. 160. K.K. Dompere
Cost-Benefit Analysis and the Theory of Fuzzy Decisions – Fuzzy Value Theory, 2004
ISBN 3-540-22161-1

Vol. 161. N. Nedjah, L. de Macedo Mourelle (Eds.)
Evolvable Machines, 2005
ISBN 3-540-22905-1

Vol. 162. N. Ichalkaranje, R. Khosla, L.C. Jain
Design of Intelligent Multi-Agent Systems, 2005
ISBN 3-540-22913-2

Vol. 163. A. Ghosh, L.C. Jain (Eds.)
Evolutionary Computation in Data Mining, 2005
ISBN 3-540-22370-3

Vol. 164. M. Nikravesh, L.A. Zadeh, J. Kacprzyk (Eds.)
Soft Computing for Information Prodessing and Analysis, 2005
ISBN 3-540-22930-2

Vol. 165. A.F. Rocha, E. Massad, A. Pereira Jr.
The Brain: From Fuzzy Arithmetic to Quantum Computing, 2005
ISBN 3-540-21858-0

Vol. 166. W.E. Hart, N. Krasnogor, J.E. Smith (Eds.)
Recent Advances in Memetic Algorithms, 2005
ISBN 3-540-22904-3

Vol. 167. Y. Jin (Ed.)
Knowledge Incorporation in Evolutionary Computation, 2005
ISBN 3-540-22902-7

Yap-Peng Tan
Kim Hui Yap
Lipo Wang (Eds.)

Intelligent Multimedia Processing with Soft Computing

Springer

Prof. Yap-Peng Tan
Nanyang Technological University
School of Electrical and
Electronic Engineering
Nanyang Avenue
Singapore 639798

Prof. Kim Hai Yap
Nanyang Technological University
School of Electrical and
Electronic Engineering
Nanyang Avenue
Singapore 639798

Prof. Lipo Wang
Nanyang Technological University
School of Electrical and
Electronic Engineering
Nanyang Avenue
Singapore 639798

ISSN 1434-9922
ISBN 3-540-23053-X Springer Berlin Heidelberg New York

Library of Congress Control Number: 2004112292

Springer is a part of Springer Science+Business Media
springeronline.com

Printed in Germany

Typesetting: data delivered by editor
Cover design: E. Kirchner, Springer-Verlag, Heidelberg
Printed on acid free paper 62/3020/M - 5 4 3 2 1 0

Preface

Soft computing represents a collection of techniques, such as neural networks, evolutionary computation, fuzzy logic, and probabilistic reasoning. As opposed to conventional "hard" computing, these techniques tolerate imprecision and uncertainty, similar to human beings. In the recent years, successful applications of these powerful methods have been published in many disciplines in numerous journals, conferences, as well as the excellent books in this book series on Studies in Fuzziness and Soft Computing.

This volume is dedicated to recent novel applications of soft computing in multimedia processing. The book is composed of 21 chapters written by experts in their respective fields, addressing various important and timely problems in multimedia computing such as content analysis, indexing and retrieval, recognition and compression, processing and filtering, etc.

In the chapter authored by Guan, Muneesawang, Lay, Amin, and Lee, a radial basis function network with Laplacian mixture model is employed to perform image and video retrieval. D. Androutsos, P. Androutsos, Plataniotis, and Venetsanopoulos investigate color image indexing and retrieval within a small-world framework. Wu and Yap develop a framework of fuzzy relevance feedback to model the uncertainty of users' subjective perception in image retrieval.

Incorporating probabilistic support vector machine and active learning, Chua and Feng present a bootstrapping framework for annotating the semantic concepts of large collections of images. Naphade and Smith expose the challenges of using a support vector machine framework to map low-level media features to high-level semantic concepts for the TREC 2002 benchmark corpus. Song, Lin, and Sun present a cross-modality autonomous learning scheme to build visual semantic models from video sequences or images obtained from the Internet. Xiong, Radhakrishnan, Divakaran, and Huang summarize and compare two of their recent frameworks based on hidden Markov model and Gaussian mixture model for detecting and recognizing "highlight" events in sports videos.

Exploiting the capability of fuzzy logic in handling ambiguous information, Ford proposes a system for detecting video shot boundaries and classifying them into categories of abrupt cut, fade-in, fade-out, and dissolve. Li, Katsaggelos, and Schuster investigate rate-distortion optimal video summarization and compression. Vigliano, Parisi, and Uncini survey some recent neural-network-based techniques for video compression. Doulamis presents an adaptive neural network scheme for segmenting and tracking video objects in stereoscopic video sequences.

Emulating the natural processes in which individuals evolve and improve themselves for the purpose of survival, Wu, Lin, and Huang propose an efficient genetic algorithm for problems with a small number of possible solutions and apply it to block-based motion estimation in video compression, automatic facial feature extraction, and watermarking performance optimization. Zhang, Li, and Wang present two recognition approaches based on manifold learning algorithm with linear discriminant analysis and nonlinear autoassociative modeling to solve the problems of face and character recognition. Chen, Er, and Wu adopt a combination of discrete cosine transform and radial basis function network to address the challenge of face recognition.

Dealing with uncertain assertions and their causal relations, Tao and Tan present a probabilistic reasoning framework to incorporate domain knowledge for monitoring people entering or leaving a closed environment. Nakamura, Yotsukura, and Morishima utilize synchronous multi-modalities, including the audio information of speech and visual information of face, for audio-visual speech recognition, synthesis, and translation. Cheung, Mak, and Kung propose a probabilistic fusion algorithm for speaker verification based on multiple samples obtained from a single source.

Er and Li develop adaptive noise cancellation using online self-enhanced fuzzy filters with applications to audio processing. Wang, Yan, and Yap propose a noisy chaotic neural network with stochastic chaotic simulated annealing to perform image denoising. Sun, Yan, and Sclabassi employ an artificial neural network to provide numerical solutions in the EEG analysis. Lienhart, Kozintsev, Budnikov, Chikalov, and Raykar present a novel setup involving a network of wireless computing platforms with audio-visual sensors and actuators, and propose algorithms that can provide both synchronized inputs/outputs and self-localization of the input/output devices in 3D space.

We would like to sincerely thank all authors and reviewers who have spent their precious times and efforts to make this book a reality. Our gratitude also goes to Professor Janusz Kacprzyk and Dr. Thomas Ditzinger for their kindest support and help with this book.

Yap-Peng Tan
Kim-Hui Yap
Lipo Wang

Singapore, July 2004

Contents

Human-Centered Computing for Image and Video Retrieval

L. Guan[1], P. Muneesawang[2], J. Lay[3], T. Amin[1], I. Lee[3]

[1]Ryerson University, Canada, [2]Naresuan University, Thailand, [3]The University of Sydney, Australia

Abstract. In this chapter, we present retrieval techniques using content-based and concept-based technologies, for digital image and video database applications. We first deal with the state-of-the-art methods in a content-based framework including: Laplacian mixture model for content characterization, nonlinear relevance feedback, combining audio and visual features for video retrieval, and designing automatic relevance feedback in distributed digital libraries. We then take an elevated post, to review the defining characteristic and usefulness of the current content-based approaches and to articulate any required extension in order to support semantic queries.

Keywords: content-based retrieval, concept-based retrieval, digital library, intelligent digital asset management, applied machine learning

1 Introduction

Content-based indexing and retrieval of multimedia data has been one of the focal research areas in the multimedia research community. Recognizing the fact that the extension of the well studied information retrieval (IR) to multimedia content is constrained by numerous limitations, the search for a better search engine started in the late 80s and early 90s of the last century. The main focus has been on indexing and retrieval of the visual data – images and videos. Early efforts for fully automated retrieval have been proven to be less effective due to two facts: (1) the representation gap between the low level features used by the computers and the high level semantics used by the humans; (2) subjective evaluation of the retrieval results. To alleviate the problems, direct participation of human users in a relevance feedback (RF) loop became a popular approach. However, the restriction of relevance feedback is obvious: excessive human subjective errors and inconvenience in networked digital libraries are two of them. Unsupervised learning have been introduced to automatically integrate human perception knowledge in order to solve the problems and preliminary results show that the approach is promising. However, the fundamental issue in retrieval cannot be completely resolved by content-based methods alone. Researchers have directed their attention to the concepts behind how human beings analyze visual scenes. Semantic integration of audio/video/text has been investigated by several groups. A more daring approach, purely based on the study of concepts of different activities, led to a novel paradigm to develop new ways of indexing and search of audio/visual documents. In this chapter, we will first survey state-of-the-art in image/video retrieval. We will then present some of our recent works in human-centered computing in content-based retrieval (CBR) and indexing audio/visual documents by concepts.

This chapter is organized as follows. Section 2 presents the methods for feature extraction and query design in relevance feedback based CBR system. Section 3 presents video retrieval using joint processing of audio and visual information. Section 4 presents new architecture of automatic relevance feedback in networked database systems. Section 5 reviews concept-based retrieval techniques.

2 Feature Extraction and Query Design in CBR

In this section, we firstly propose a Laplacian mixture model (LMM) for content characterization of images in the wavelet domain. Specifically, the LMM is used to model the peaky distributions of the wavelet coefficients. It extracts a low dimensional feature vector which is very important for the retrieval efficiency. We then study a non-linear approach for similarity matching within the relevance feedback framework. An adaptive radial basis function network (ARBFN) is proposed for the local approximation of the image similarity function. This learning strategy involves both positive and negative training samples. Thus, the current system is capable in modeling user response with minimum feedback cycles and a small number of feedback samples.

2.1 Feature Extraction

The wavelet transform of most of the signals we come across in the real world are sparse due to its compression property. There are a few wavelet coefficients that have large values and carry most of the information, while most of the coefficients are small. This energy packing property of the wavelet coefficients results in a peaky distribution. This type of peaky distribution is more heavy-tailed than the Gaussian distribution. In Figure 1, we have plotted the histograms of wavelet coefficients at different scales for an example image from the Brodatz image database. The peaky nature of the distributions is clearly observed from this figure.

As illustrated above, the distributions of the wavelet coefficients are non-Gaussian in nature. Therefore modeling of wavelet coefficients using a single

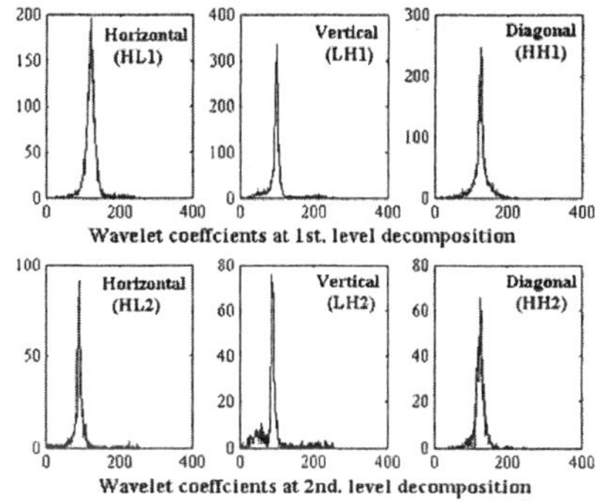

Fig. 1. Histograms of wavelet coefficients at different scales of a texture image from the Brodatz image database.

distribution such as Gaussian or Laplacian gives rise to mismatches. The mixture modeling provides an excellent and flexible alternative for this kind of complex distribution. Finite mixture models (FMMs) are widely used in the statistical modeling of data. They are a very powerful tool for probabilistic modeling of the data produced by a set of alternative sources. Finite mixtures represent a formal approach to unsupervised classification in statistical pattern recognition. The usefulness of this modeling approach is not limited to clustering. FMMs are also able to represent arbitrarily complex probability density functions [1].

We can model any arbitrary shaped distribution using mixture of Gaussians if we have an infinite number of components in the mixture. This is however practically infeasible. We therefore model the wavelet coefficient distribution with a two component Laplacian mixture. The parameters of this mixture model are used as features for indexing the texture images. It has been observed that the resulting features possess high discriminatory power for texture classification. Because of the low dimensionality of the resulting feature vector, the retrieval stage consumes less time enhancing the user experience while interacting with the system. The images are decomposed using 2-dimensional wavelet transform. The 2-D wavelet transform decomposes the images into 4 subbands representing the horizontal, vertical, diagonal information and a scaled down low resolution approximation of the original image at the coarsest level. The texture information is carried by only a few coefficients in the wavelet domain where the edges occur in the original images.

In our method, we model the wavelet coefficients in each wavelet subband as a mixture of two Laplacians centered at zero:

$$p\left(w_i\right) = \alpha_1 p_1\left(w_i|b_1\right) + \alpha_2 p_2\left(w_i|b_2\right), \quad \alpha_1 + \alpha_2 = 1 \tag{1}$$

where α_1 and α_2 are the mixing probabilities of the two components p_1 and p_2; w_i are the wavelet coefficients; b_1 and b_2 are the parameters of the Laplacian distribution p_1 and p_2 respectively. The Laplacian component corresponding to the class of small coefficients has relatively small value of parameter b_1. The Laplacian component in (1) is defined as:

$$p_1\left(w_i|b_1\right) = \frac{1}{2b_1}\exp\left(-|w_i|/b_1\right) \tag{2}$$

The shape of the Laplacian distribution is determined by the single parameter b.

We apply the EM algorithm [2] to estimate the parameters of the model. The EM algorithm is iterative and consists of two steps, E-step and M-step, for each iteration.

E-Step: For the *n-th* iterative cycle, the E-step computes two probabilities for each wavelet coefficient:

$$p_{1i}(n) = \frac{\alpha_1\left(n\right)p\left(w_i|b_1\left(n\right)\right)}{\alpha_1\left(n\right)p\left(w_i|b_1\left(n\right)\right) + \alpha_2\left(n\right)p\left(w_i|b_2\left(n\right)\right)}, \tag{3}$$

$$p_{2i}(n) = \frac{\alpha_2\left(n\right)p\left(w_i|b_2\left(n\right)\right)}{\alpha_1\left(n\right)p\left(w_i|b_1\left(n\right)\right) + \alpha_2\left(n\right)p\left(w_i|b_2\left(n\right)\right)} \tag{4}$$

M-Step: In the M-step, the parameters $[b_1, b_2]$ and *a priori* probabilities $[\alpha_1,\ \alpha_2]$ are updated.

$$\alpha_1\left(n+1\right) = \frac{1}{K}\sum_{i=1}^{K} p_{1i}\left(n\right), \quad \alpha_2\left(n+1\right) = \frac{1}{K}\sum_{i=1}^{K} p_{2i}\left(n\right), \tag{5}$$

$$b_1\left(n+1\right) = \frac{\sum_{i=1}^{K}|w_i|p_{1i}\left(n\right)}{K\alpha_1\left(n+1\right)}, \quad b_2\left(n+1\right) = \frac{\sum_{i=1}^{K}|w_i|p_{2i}\left(n\right)}{K\alpha_2\left(n+1\right)} \tag{6}$$

where K is the total number of wavelet coefficients.

To obtain the content features, an image is firstly decomposed using 2-D wavelet transformation. The EM algorithm is then applied to each of the detailed sub-bands LH, HL, HH at each wavelet scale. The model parameters $[b_1, b_2]$ calculated for each subband are used as features. The mean and standard deviation of the wavelet coefficients in the approximate subband are also chosen as features. In case of 3-level decomposition of images, the feature vector is 20-dimensional. The individual components of the feature vector have different dynamic ranges because they measure the different physical quantities. Therefore the feature values are rescaled to contribute equally to the distance calculation.

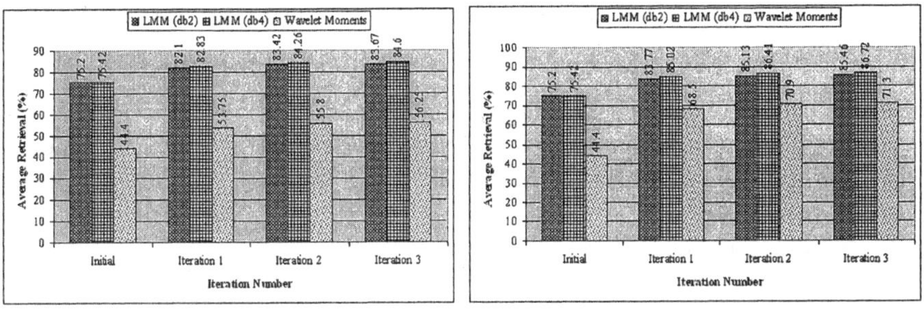

(a) (b)

Fig. 2. Average Recall (%) obtained by retrieving 1856 query images using (a) query modification approach; and (b) RBF approach. In both cases, the initial results are based on the city-block distance.

Retrieval Performance

The discriminatory power of the features is highly important for an effective image retrieval system. However, it is very difficult to model the human visual perception by only a set of features. Also the similarity between the images is a very subjective notion. The visual content of the images may be interpreted differently by different individuals. The objective of an efficient CBR system is to model human visual system. This serves as the motivation for the idea of relevance feedback (RF). Relevance feedback is a mechanism of learning from user interaction. The system parameters are changed depending on the feedback from the user. There may be a variety of ways in which the input from the user can be used. In our experiments, a query modification (QM) approach along with a single-class radial basis function (RBF) for similarity criteria is employed [3].

Figure 2(a) summarizes the retrieval performance of the proposed feature set using query modification approach in the RF. These experimental results were obtained using Brodatz texture image database. Brodatz image database contains 1856 images divided into 116 classes. Every class contains 16 images. It is observed that Laplacian Mixture Model (LMM) features perform significantly higher than Wavelet Moments (WM). A performance increase of 31.02 % is achieved in the initial search cycle. The retrieval ratio of 84.60 % is obtained at third iteration compared to 56.25 % in WM case.

Figure 2(b) depicts the performance of RBF for both feature sets. An increase of 2.12 % is obtained for the LMM features compared to 15.05 % increase in case of WM features. It is further observed the performance is slightly higher in case the images are decomposed using Daubechies-4 (db4) wavelet kernel compared to Daubechies-2 (db2).

2.2 An Adaptive Radial Basis Function Network for Query Modeling

In order to learn user perception through a relevance feedback process, we propose an adaptive radial basis function network (ARBFN) for query modeling using multiple-modeling paradigm. In this framework, a function approximation associated with a given query is estimated by the superposition of different local models. Via the three-layer architecture of the RBF network, the discriminant function is obtained by a linear combiner as:

$$f(\mathbf{x}) = \sum_{i=1}^{n_h} \Phi_i \left(||\mathbf{x} - \mathbf{c}_i|| \right) \cdot \theta_i \tag{7}$$

where $\mathbf{x} \in \Re^P$ denotes an input vector; $\Phi_i(\cdot)$ is the nonlinear model function; $\mathbf{c}_i \in \Re^P$ and θ_i are the corresponding RBF center and linear weight, respectively. The advantage of this network used in the current application is that it finds the input–output map using local approximators. Consequently, the underlying basis function responds only to a small region of the input space where the function is centered, e.g., a Gaussian response, $\Phi(\gamma) = e^{-(\gamma^2/\sigma^2)}$, where σ is a real constant, and $\Phi(\gamma)= 0$ as $\gamma \to \infty$. This relationship allows *local evaluation* for image similarity matching.

Unfortunately, due to the possible high correlation between training samples introduced during relevance feedback process, the general criteria previously studied (e.g., [4,5]) to select $\mathbf{c}$ and σ will not guarantee adequate performance. The uniqueness of the image retrieval application introduces new challenges in the construction of the RBF model. A small training set feedback by the user during interactive cycle contains samples that are highly correlated to each other. This correlation is in terms of visual similarity as well as numerical distance in feature space. The EDLS (Exact Design Network using Least Square criterion) [4] provides us with a useful example of the problem of numerical ill-conditioning which is caused by some centers being too close to each other or highly correlated. This is due to the fact that the EDLS derives RBF centers from all training samples in a one-to-one corresponding manner. Chen's original orthogonal least squares (OLS) algorithm [5] to select a possible subset of samples for RBF centers so that adequate and parsimonious RBF networks can be derived. The OLS method is employed as a forward regression procedure by treating the centers as the regressors, and selecting a subset of significant regressors from a given candidate set. This regression procedure also allows monitoring regressors that cause numerical ill-conditioning. However, in the image retrieval application, the criterion for selecting RBF centers employed by OLS may not adequately address the high level of correlation among training samples.

Network training

Within a feedback cycle, we may form a training sample set for the RBF network as: $T = \{\mathbf{x}_1, \mathbf{x}_2, ..., \mathbf{x}_N\}$, each sample having distance, $D_i^{(\mathbf{z}_j)}$, with

$$D_1^{(\mathbf{z}_j)} \leq D_2^{(\mathbf{z}_j)} \leq \cdots \leq D_N^{(\mathbf{z}_j)} \tag{8}$$

where $D_i^{(\mathbf{z}_j)}$ denotes the distance between $\mathbf{x}_i$ and a query $\mathbf{z}_j$, chosen from the N_T data points in the entire database, and $N << N_T$. Each data point in this training set is labeled either as positive or negative. We formally denote the positive sample set as $\mathrm{X}^+ = \{(\mathbf{x'}_i, y_i) \mid y_i = 1,\ i = 1, ..., n_M\}$, and the negative sample set as $\mathrm{X}^- = \{(\mathbf{x''}_i, y_i) \mid y_i = 0,\ i = 1, ..., n_N\}$, where n_M and n_N are the total number of positive and negative samples, respectively.

Due to the small size of set T with the high correlation between its members, we proposed a new ARBFN learning algorithm to characterized RBF network model. The proposed learning strategy for RBF network involves two phases. First, the positive feedback strategy is used to construct the local approximators and their associated centers and widths. Second, the negative feedback strategy is used to improve decision boundary of the network.

Estimation of Approximators via Positive Feedback

In general, both positive and negative samples may be chosen as the RBF centers. However, the application of negative samples to RBF centers and their associated linear weights may reduce classification power of the RBF network. This is due to the small distance among the retrieved samples as described by (8), which make the selected centers too close. A possible solution to overcome this problem is to assign the RBF centers only with positive samples. By this way, the shape of each positive cluster can be described by:

$$\Phi_i\left(||\mathbf{x} - \mathbf{c}_i||\right) = \exp\left(-\frac{||\mathbf{x} - \mathbf{c}_i||^2}{2\sigma_i^2}\right) \tag{9}$$

where $\mathbf{c}_i = \mathbf{x'}_i,\ i = 1, ..., n_M$. In addition, as oppose to the OLS algorithm, each RBF width σ in (9) is adjusted dynamically according to the distance between RBF center and its neighbor. That is:

$$\sigma_i = \zeta \cdot \min_j\left(||\mathbf{c}_i - \mathbf{c}_j||\right),\ \forall\{i,j\} \in \{1, ..., n_M\},\ i \neq j \tag{10}$$

where ζ is an overlapping factor.

The most accurate means of obtaining the linear weights, $\theta_i, i = 1, ..., n_M$, comes from applying the least squares algorithm to the RBF model [4]. This algorithm produces the solution to minimizing least square errors as:

$$\mathbf{\Theta} := [\theta_1, \theta_2, ..., \theta_{n_M}]^T = \left(\mathbf{G}^T\mathbf{G}\right)^{-1}\mathbf{G}^T\mathbf{y} \tag{11}$$

where $\mathbf{y} = [y_1, y_2, ..., y_{n_M}]^T$ is the decided response, and the matrix $\mathbf{G} \in \mathrm{M}_{n_M \times n_M}$,

$$\mathbf{G} = \{\Phi_{ij}\},\ \Phi_{ij} = \exp\left(-\frac{||\mathbf{x'}_i - \mathbf{c}_j||^2}{2\sigma_j^2}\right), \quad i, j = 1, .., n_M \tag{12}$$

In our work, since only positive samples are chosen as RBF centers, together with $y_i = 1, \forall i$, we set $\theta_i = 1,\ \forall i = 1, ..., n_M$, without the application of (11)-(12). This also provides simplicity for practical use of the RBF network.

The basic RBF version of the ARBFN discussed in (9) is based on the assumption that the feature space is uniformly weighted in all directions. However, image feature variables tend to exhibit different degrees of importance which heavily depend on the nature of the query and the relevant images defined [3]. This leads to the adoption of an elliptic basis function (EBF). Thus, in place of (9), a new version of the Gaussian-shaped RBF that takes into account the feature relevancy, can be defined as:

$$\Phi_i\left(||\mathbf{x} - \mathbf{c}_i||_A\right) = \exp\left(-\frac{(\mathbf{x} - \mathbf{c}_i)^T \boldsymbol{\Lambda} (\mathbf{x} - \mathbf{c}_i)}{2\sigma_i^2}\right) \tag{13}$$

where $\boldsymbol{\Lambda} = diag[\alpha_1, ..., \alpha_p, ..., \alpha_P]$. So, the parameters $\alpha_p,\ p = 1, ..., P$ represent the relevance weights which are derived from the variance of the positive samples in X^+, i.e., $\alpha_p \propto 1/\xi_p$, where ξ_p is the standard deviation of the p-th feature variable of $\mathbf{x'}_i, i = 1, ..., n_M$.

Expansion Centers via Negative Feedback

The possibility of moving the expansion centers is useful for improving the representativeness of the centers. Recall that, in a given training set, we have both positive and negative samples, which are the ranked results from the previous search operation. For the negative samples, $\mathbf{x''}_i,\ i = 1, ..., n_N$, the similarity scores from the previous search indicate that their clusters are close to the positive samples retrieved in the same relevance feedback cycle. Here, the use of negative samples becomes essential, as the RBF centers should be moved slightly away from these clusters. Therefore, we modify each positive sample $\mathbf{x'}_i,\ i = 1, ..., n_M$, before applying it to characterize the RBF centers in (9). The modification involves on iteration process of the following two steps:

- Step I, select a data point $\mathbf{x''}$ from $\{\mathbf{x''}_i,\ i = 1, ..., n_N\}$, and choose the winning node $\mathbf{x'}_{i^*}$, according to:

$$i^* = \arg \min_{i \in \{1,...,n_M\}} \left((\mathbf{x''} - \mathbf{x'}_i)^T \boldsymbol{\Lambda} (\mathbf{x''} - \mathbf{x'}_i)\right) \tag{14}$$

- Step II, modify the winning node by the anti-reinforced learning rule,

$$\mathbf{x}'_{i^*}(t+1) = \mathbf{x}'_{i^*}(t) - \beta[\mathbf{x}''(t) - \mathbf{x}'_{i^*}(t)] \tag{15}$$

  and decrease the learning constant $\beta(t)$ monotonically, $\beta(t) = \beta(t)/(\beta(t) + 0.5)$.
- Repeat Step I and Step II until $\beta(t) \approx 0$.

Retrieval Performance

In this section we demonstrate retrieval performance of the ARBFN method by comparing it with the EDLS [4] and OLS [5] algorithms. The methods were applied to the image database from Corel Gallery 65000 product [6], which contains 40,000 photographs from 400 categories. Each image was indexed by multiple types of image features, including color histogram and color moments for color descriptors; Gabor wavelet transform for texture descriptors; and Fourier descriptor for shape descriptors. Since these images were originally organized by Corel professionals that provide us the ground truth, PR[1] results serve as a meaningful indication of retrieval accuracy.

The proposed ARBFN algorithm was applied for retrieval using its criterion for selection of RBF center and width, together with the shifting center. In the comparing method, the OLS learning procedure was used to choose the RBF centers as a subset of training data vectors, using a one-by-one selection manner and error reduction criterion. The selection process is terminated when the network's mean squared error falls below a pre-defined tolerance. In comparison, the EDLS assigned all feedback samples to the RBF centers. EDLS constructed the second layer weights and bias in such a way that the network sum-squred error was minimized to zero on the training vectors. The RBF widths used by OLS and EDLS were determined experimentally, and it was found that the appropriate width was $\sigma = 0.8$.

Table 1 summaries the average precision results, $\overline{\text{PR}}(rf)$, as a function of iteration rf, taken over 35 queries. We can see from the results that the ARBFN significantly improved the retrieval accuracy (up to 92% precision). The first iteration showed an improvement of about 35.9% ; the second iteration an additional 9.6% ; and the third increased by 2.1% . The ARBFN outperformed the OLS (76.61%) and the EDLS. This result confirms that the ARBFN learning strategy offers a better solution to the construction of a RBF network for the interactive image retrieval, than the two standard learning strategies. In addition, it was observed that the performance of the EDLS reduced after two iterations as the retrieved samples become correlated more strongly. This indicated that the RBF centers critically influenced the performance of the RBF classifier, and that RBF classifier constructed by matching all retrieved samples exactly to the RBF centers degraded the retrieval performance. Based on our simulation study, it is evident that the ARBFN is very

[1]PR (precision rate) is defined as number of relevant image retrieved over top sixteen best matches.

effective in learning from a given small set of feedback samples, and converges quickly within one to two feedback cycles.

Table 1. Average precision (%) as a function of relevance feedback cycles, $\overline{\text{PR}}(rf)$, obtained by retrieving 35 quires, using Corel Database.

Method	**Average Precision (%), $\overline{\text{PR}}(rf)$**			
	rf=0	*rf=1*	*rf=2*	*rf=3*
ARBFN	44.82	80.72	90.36	92.50
EDLS	44.82	50.18	43.39	43.04
OLS	44.82	66.07	73.21	76.61

3 Audio Visual Cues for Video Indexing and Retrieval

In this section we describe a video indexing and retrieval technique using combined processing of audio and visual information. Compared to still images, the spatio-temporal information of video file contains multimodality signal fusion of image sequence, music, caption text, and spoken words. When used together, the multi-modality signals form powerful features for analyzing video content. In sport video, audiovisual analysis is effective to detect sport key episodes, such as goal event, touchdown sequence [7], and racket hits in a tennis game [8]. In broadcast television, multimodality signals can be used to classify video into certain types of scene like dialogue, story, action [9], weather forecasts, and commercials [10]. Multimodality also plays an important role in combining content and context features for multimedia understanding which involves multimedia object detection and organization [11,12]. This technique aims at supporting a better interface of high-level-concept queries.

Unlike the previous works discussed, the proposed method owns its scalability and flexibility properties. Firstly, the scalability demonstrates that the current technique can be applied to a wide range of applications. As compared to the techniques in [7-9], the proposed method does not restrict itself either to specific video domains or pre-defined event and scene classes. Secondly, the proposed technique is flexible in that it can be applied to a longer portion of video clip, beyond the shot or key frames. We adopt *adaptive video indexing* (AVI) technique to characterize visual content. AVI can be applied to shot, scene, and story levels [13]. Also, the content analysis of audio is obtained by statistical time-frequency analysis methods that can be applied to audio at the clip level. This method is independent to the pre-defined audio segments. This means that the proposed method assumes the audio as a non-stationary signal, in which the signal characteristic can be changed dramatically within a given portion of audio clip.

3.1 Visual Modeling by Adaptive Video Indexing (AVI) Technique

Our motivation for visual content characterization is based on the fact that video data is a sequence of images. Videos which have similar contents usually contain similar images. The degree of similarity between videos may depend on the degree of "overlapping", describing how often the videos refer to a similar set of images. In general, a primary content description for video interval I_x can be defined by:

$$D_{I_x} = \{(X_i, f_i) \mid X_i \in \Re^p,\ i = 1, 2, ..., n_f\} \tag{16}$$

where f_i denotes the i-th video frame, X_i is its feature vector; n_f is the total number of frames within I_x; and p is the dimension of X_i. It is noted that the video interval I_x can be of any levels as defined by shot, scene, and story clips. These organizations facilitate multiple-level access to video databases [13].

For video indexing, the key-frame based method is commonly used to model the video content within shot level by the selection of optimum frames. However, this need not necessarily be the case for the video characterization scenario, with which we are ultimately concerned, to address the temporal information. Smith *et al* [14] approached this problem by modeling video content along temporal dimension. The model matches descriptors for multiple frames from the query and target video. This scenario is employed by the proposed AVI method to take into account all frames along temporal dimension. However, instead of directly storing and matching all X_i as in Smith's model [14], the AVI defines a descriptor to wrap spatio-temporal information using a probability feature transformation.

With the AVI model, a descriptor is defined by the probability of finding a visual model (or template) M_t in the input video, which is given simply by:

$$P(M_t) = (n_f \times \eta)^{-1} \sum_{j=1}^{n_f \times \eta} I(\ell_j = \ell_{M_t}) \tag{17}$$

where ℓ_{M_t} is the label of the corresponding model vector M_t, and $n_f \times \eta$ is the total number of labels used for indexing the input video. The function $I(\cdot)$ is equal to 1 if the argument is true, and 0 otherwise. The descriptor $P(M_t)$ is estimated for its corresponding model vector in the model set, $\mathbf{M} = [M_1, ..., M_t, ..., M_T]$, $M_t \in \Re^p$. These models are generated and optimized from the training vectors X_j, $j = 1, ..., J$. Here, we shall assume that the number of model vectors is significantly smaller than the number of training vectors, i.e., $T << J$. In order to obtain a model vector set, $\mathbf{M}$ we apply the competitive learning algorithm [15] to the training set X_j, $j = 1, ..., J$. Here, the space of X is characterized by a color histogram feature, using HSV color space and 48 bins, i.e., $p = 48$.

For video indexing, a secondary descriptor is generated to characterize spatio-temporal information. We generate a set of labels via a multiple-label-mapping function $F(X) : \Re^{48} \rightarrow \Re^{\eta}$, where each X_i is mapped onto a Voronoi space through:

$$X_i \Rightarrow \langle t^*, R^{\eta}_{t^*} \rangle \Rightarrow \rho^{(X_i)}, \quad t^* = \underset{t \in \{1,\dots,T\}}{\arg\min} \left(\|X_i - M_t\| \right), \quad R^{\eta}_{t^*} = \cup^{\eta}_{j=1} M_j, \tag{18}$$

where $R^{\eta}_{t^*}$ is a region containing η Voronoi cells—the wining node M_{t^*} and other $\eta - 1$ Voronoi cells neighboring to the node M_{t^*}. The set of labels, $\rho^{(X_i)} = \ell^{X_i}_{t^*,1}, \ell^{X_i}_{t^*,2}, \dots, \ell^{X_i}_{t^*,\eta}$, contains the associated labels of the Voronoi cells in $R^{\eta}_{t^*}$. In other words, all labels, $\ell^{X_i}_{t^*,j}$, $j = 1,..,\eta$, represent the top η best match models to the input vector X_i. This multiple-label mapping process allows the interpretation of the correlation information among the models. It also allows the interpretation of the correlation information between video frames for better analysis, since these frames are usually highly correlated in the input video.

The mapping of all video frames in I_x results in a set of $\rho^{(X_i)}$, $i = 1, 2, .., n_f$, which is then concatenated into a single feature vector $W_{I_x} = [w_1, \dots, w_t, \dots w_T]$. According to (17), each w_t is obtained by:

$$w_t = P(M_t) = (n_f \times \eta)^{-1} \sum_{i=1}^{n_f} \sum_{j=1}^{\eta} I(\ell^{X_i}_j = \ell_{M_t}). \tag{19}$$

It is noted that only a few of w_t have nonzero value since $T >> n$. Therefore, W is very sparse, which allows for efficient storage space and fast vector matching. This sparse vector can be formally defined as:

$$\hat{W}_{I_x} = \left[(\hat{t}, w_{\hat{t}}) \right], \; \hat{t} \in \chi, \; \chi \subset \{1, 2, \dots, T\}, \tag{20}$$

where χ is a set of indexes t belonging to the nonzero w_t. In this way, the content similarity matching between two video intervals, I_x and I_y can be computed by:

$$S_{xy} = 1 - \hat{W}_{I_x} \cdot \hat{W}_{I_y} \Big/ \left(||\hat{W}_{I_x}|| \times ||\hat{W}_{I_y}|| \right) \tag{21}$$

where the second term is defined by the popular cosine measure.

3.2 Feature Extraction from Embedded Audio

Audio is a rich source of information that often reflects what is happening in the video scenes. The users of the video data might be interested in certain action sequences that are easier to identify in the audio domain. Visual information may not yield useful indices in this kind of scenario. For this purpose, we need to extract features that represent the global similarity of the embedded audio content. A statistical approach has been adopted here to analyze the audio data and extract features for video indexing.

The proposed indexing scheme does not depend on the segmentation method. The video may be segmented into clips using any existing algorithm. The embedded audio is separated from the video clips, and then re-sampled at a uniform sampling rate. Each audio segment is decomposed using a one-dimensional Discrete Wavelet Transform (DWT) which decomposes the signal into 2 subbands at each wavelet scale; a low frequency subband and a high frequency subband. Audio signal is different from image signal. In images, the values of adjacent pixels usually don't change sharply. On the other hand digital audio signal is a form of oscillating waveform, which includes a variety of frequency components varying with time. Most audio signals consist of a wide variety of frequencies. The wavelet coefficients of audio signal have many large values in detail levels, and the lowest frequency subband coefficients do not always provide good approximation of the original signal. The wavelet decomposition scheme matches the models of sound octave-division for perceptual scales. Wavelet transform also provides a multi-scale representation of sound information, so that we can build indexing structure based on this scale property. Moreover, audio signals are non-stationary signals whose frequency contents evolve with time. Wavelet transform provides both frequency and time information simultaneously. These properties of wavelet transform for sound signal decomposition forms the foundation of the audio retrieval and indexing system developed in this section.

The wavelet decomposition is taken up to 9 levels. An increase in the level of decomposition increases the number of features extracted for indexing. This improves the retrieval performance at the expense of more computational overhead. The statistical model based on the Laplacian mixture of two components developed for the texture retrieval in Section 2.1 is applied for feature extraction from the embedded audio. It is observed that parameters of the model are a good representation of the audio segments and define the global characteristics of the audio. The parameters for this statistical model are estimated using the EM algorithm.

The following components form the feature vector used for indexing the video clips: (a) Mean and standard deviation of the wavelet coefficients in the Low frequency subband, (b) Model parameters $[\alpha_1, b_1, b_2]$ calculated for each of the high frequency subband (cf. Eqs. (5)-(6)). The feature vector is 29-dimensional in case of 9-level wavelet decomposition of the audio clips. The components of the feature vector represent different physical quantities and have different dynamic ranges. Therefore normalization of the feature vector is required to put equal emphasis on each component.

3.3 Combining Audio and Visual Information

The audio and visual features may not be directly combined because they are different in both physical structure as well as their feature dimensions. Thus, the video ranking is obtained separately from audio and visual feature databases. These ranking results are then combined to obtain the final similarity ranking decision. Using the visual feature database and (21), the similarity scores between the query interval I_q and other video intervals in the database are generated. This results in $S_{qi}^{(v)}$, $i = 1, ..., N_T$, where N_T denotes the total number of video files in the database. These scores are sorted in increasing order, so that video interval I_i can be associated with its ranking index, $Rank_{I_i}^{(v)}$. In this way, $Rank_{I_i}^{(v)} < Rank_{I_j}^{(v)}$ if $S_{qi}^{(v)} < S_{qj}^{(v)}$, $\forall j \neq i$. Similarly, the ranking of audio feature database produces the similarity scores, $S_{qi}^{(a)}$, $i = 1, ..., N_T$, which are used to obtained ranking indexes, $Rank_{I_i}^{(a)}$, $i = 1, ..., N_T$. For the i-th video interval I_i, the resulting ranking scores from visual and audio, $Rank_{I_i}^{(v)}$ and $Rank_{I_i}^{(a)}$ are then combined to obtain a new similarity score as:

$$S_{qi}^{(v,a)} = Rank_{I_i}^{(v)} + (\xi) Rank_{I_i}^{(a)}, \qquad (22)$$

where ξ is the scaling factor, which is set to $0 \leq \xi \leq 1$, so as to control the impact of audio-feature ranking outcomes. This combination produces a new set of similarity scores, $S_{qi}^{(v,a)}$, $i = 1, ..., N_T$ which are arranged to obtain the retrieval set. It is noted that the selection criteria for ξ in (22) is dependent on the application. In general, the number of classes defined by the audio features is usually small compared to the one defined by visual features. In the experiment, we see that the current system performed well with the proper choice of parameter ξ, while its performance is decreased when $\xi \to 1$.

3.4 Experimental Results

For our video database, we use three Hollywood movies, fifteen music videos, and six commercial videos. These videos are segmented into 6,000 video clips, each of which contains one to three shots, and has length of approximately 30 seconds. For visual descriptor, we choose a model set length of $T = 2000$, which is used to index video and obtain nonzero feature vector $\hat{W}_{I_i}$, $i = 1, ..., 6000$. The feature dimension of $\hat{W}$ is in the range between 5 and 159, with the mean value $\approx$ 29.The audio feature extraction discussed in Section 3.2 was applied to the video database, where a 29-dimensional vector is used as an audio feature. Figure 3 shows the GUI and query interface implemented by this system.

Twenty five queries were generated from different high-level query concepts that include “fighting”, “ship crashing”, “love scene”, “music video”, and “dance party”. We used five queries for each concept, and measured retrieval precisions from the top 16 best matches. Figure 4 compares precision

Fig. 3. GUI and query interface of the proposed CBVR system using audio-visual descriptors.

results obtained by using audio descriptor, visual descriptor, and audio-visual descriptor. The results in this figure are obtained by averaging the precisions within the query concepts, as well as within the overall queries. This figure clearly reveals the benefits of combined audio-visual features. Using the visual and audio features together yielded the highest retrieval accuracy at 94.8% precisions. These results show that the audio and visual features are robust to the nature of query. This system takes advantage of the dominant feature in each query concept to achieve high retrieval accuracy, which indicates the importance of using multimodality signal analysis for video retrieval.

4 Image Retrieval in Distributed Digital Libraries

In the internet era, rich multimedia information is easily accessible by everyone attached to the network. While the amount of multimedia content keeps growing, locating the relevant information becomes more and more time consuming. A successful networked CBR system should therefore maximize the retrieval precision while optimizing the network resources. Most of the CBR systems proposed today assume a centralized query server, such as QBIC [16] and VisualSEEk [17]. Content query over distributed peer-to-peer (P2P) network is studied in [18], with the assumption that a peer contains only one image category. To handle the search across a substantial number of distributed

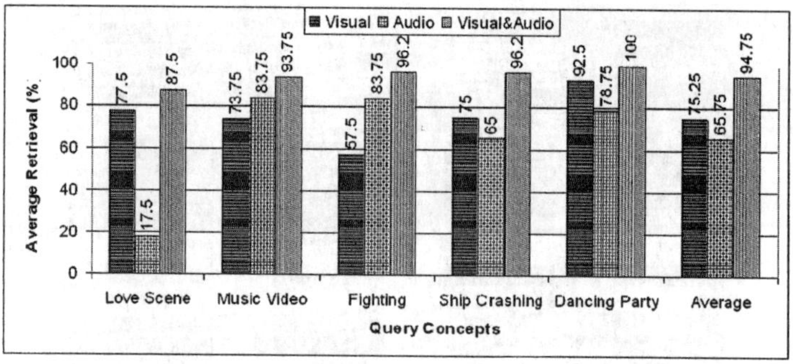

Fig. 4. Average Precision Rate (%) obtained by using audio, visual, and audio-visual descriptors, measured from five high-level query concepts.

databases, a meta-search engine (MSE) incorporating the search-agent processing model (SAPM) is proposed [19]. Here, we initiate a different approach by studying practical scenarios where multiple image categories exist in each individual database in the distributed storage network. In this scenario, we apply an automatic relevance feedback method described in [33] to the networked databases, where the system can achieve high retrieve accuracy as well as minimize network resources. The proposed system employs unsupervised learning method, the self-organizing tree map (SOTM) algorithm to minimize user interaction from the conventional relevance feedback learning process.

4.1 CBR over Centralized, Clustered, and Distributed Storage Networks

According to the distribution of the feature database, we classify CBR systems into centralized, clustered, and distributed P2P systems, as illustrated in Figure 5.

Centralized CBR System

A centralized CBR system maintains a central server to handle the query requests. Upon retrieving the relevant images according to the feature similarity measure, the universal content locator (URL) will be returned to the requesting host. The actual content will be transferred directly from the content server to the requesting host. The centralized CBR systems keep the entire feature descriptor database in a centralized server. The real image content may or may not be located on the same server. The centralized CBR server retrieves relevant content based on the feature-descriptor database. The drawback of the centralized CBR system is the scalability to handle growing retrieval requests and larger image databases.

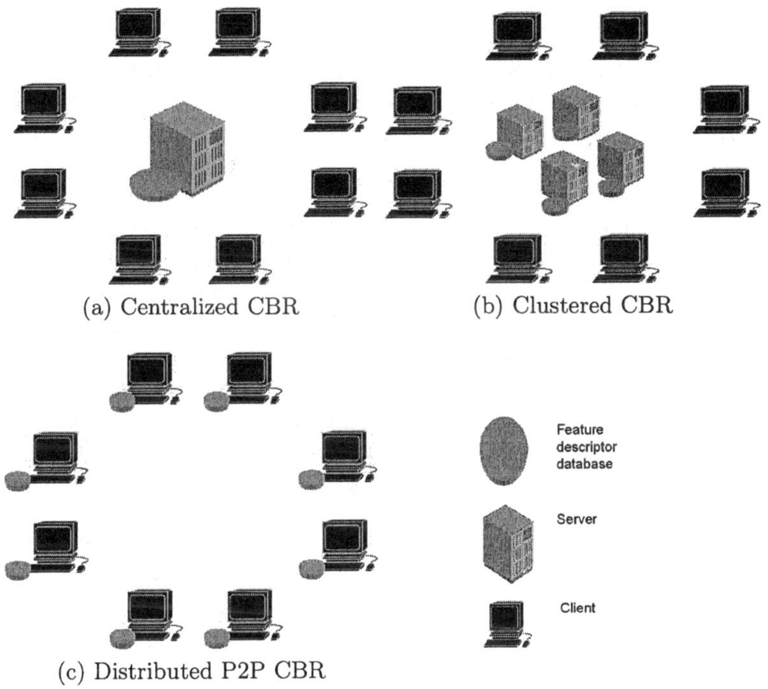

Fig. 5. Networked CBR system.

Clustered CBR System

When the image database grows, the centralized CBR approach does not scale to handle the growing computational requirement. Therefore, we propose a clustered CBR system which pre-classifies the feature descriptor database into non-overlapping categories, and each category is stored in a different server. The clustered CBR approach helps to reduce the computational and bandwidth requirement for the centralized CBR system. To address the issue of scalability for centralized CBR system, the feature descriptor database can be pre-clustered and stored in different servers. Each server in the cluster will pre-compute the centroid, which is the mean of the feature descriptors stored in this particular server.

During the query stage, the best query server is firstly identified from the similarity measure of the query feature descriptor and the cluster centroid using the nearest neighbor algorithm. Once the query server is identified, relevant content will be retrieved within the feature descriptor database stored on this server.

In our study, only the best server is identified to perform the content query. Content query over multiple best-matched servers may potentially result in better retrieval precision at the cost of computation and bandwidth.

4.2 Distributed CBR System

To further acknowledge the high correlation between each individual image database resulting from, for example, hobbyist photo collections, a decentralized CBR system attempts to group distributed nodes which share the same image categories. Then, retrievals are made at each node. A special case of the decentralized database system using the peer-to-peer (P2P) network is studied. Each node in the P2P network acts both as a client for requesting images and a server for re-distributing the images. Since a peer can join and leave the network at any time, a challenge of this distributed CBR system is to address the non-guaranteed level of service of the P2P network. To localize the search, the query packet is always associated with certain Time-to-live (TTL) levels. Database storage on distributed servers had been applied in the industry to provide high availability (providing continuous service if one or more servers are unintentionally out of service) and efficiency (access from the closest server geographically). P2P network is a special case of such a network where each node in the network behaves as a database server.

The underlying assumption of the multimedia content collection on a P2P node consists of limited categories, as the user behavior determines the data collection. Distributed CBR systems can benefit from such high correlation among certain peers, hence reduce the computation and bandwidth cost by searching within a limited subset of peers. Each peer's image collection can be considered as a subset of the whole image database, and no assumption is made on the inter-dependencies or collaborating work between any two peers. Therefore, the overlap between peers can be used to improve the retrieval precision.

Each peer in the P2P CBR system maintains two tables of neighbors. The first type of neighbors are called the *generic neighbor* which typically represent the neighbors with the least physical hop counts. The other type of neighbors are called the *community neighbors* and common interest is shared among the community. Two stages of operation are required: *community neighborhood discovery* and *query within the community neighborhood.*

Community Neighborhood Discovery

As shown in Figure 6(a), a peer node originates the query request to its generic neighbors in the P2P network. Whenever a peer node receives a query request, it will (1) decrement the TTL, and forward the request to the generic neighbors when TTL > 1, and (2) perform the content search within the peer's feature descriptor database. The retrieval results of each peer are transmitted to the original query peer directly in order to improve the efficiency.

Like most P2P applications, the proposed distributed CBR system applies an application layer protocol, such that the system can be realized on today's Internet without modifying the underlying network infrastructure. The proposed query packet format to traverse through the P2P network is shown in Figure 6(c).

Once the destination peer receives the query and performs feature match, it will issue a Query Reply to the query requester directly. The query results are in the form of filenames and distance. The actual file transfer is not part of the protocol, and protocols like HTTP, RTP, with or without encryption, may be applied depending on the application. Transferring the actual image content is coupled with the feature descriptor transmission, to eliminate the need to re-compute the feature descriptors upon receiving a new image. The proposed query search and query response packet format to traverse through the P2P network are shown in Figures 6(b) and 6(d), respectively.

The query peer maintains a table of community neighbors based on past retrieval results to identify the peers which collects similar image database.

Query within the Community Neighborhood

Once the community neighbors are identified, subsequent queries will be made to limited peers within the community neighborhood. To improve the communication efficiency, instead of forwarding the request hop-by-hop in the community neighborhood discovery stage, direct communication between the peers is applied. The same packet format is used for Query and Query Response within the community neighborhood.

Each peer in the community neighborhood collects more than one category of images, with at least one common category as the requesting peer to satisfying the criteria to be listed in the community neighborhood. Therefore, the same image appears in multiple peers are likely belong to the common category in the community neighborhood. Let $Ret(I,P_n)$ denote the retrieved result using query image I from peer P_n, where $P_n \in$ {community neighborhood}. Let $N(\cap Ret(I,Pn))$ denote the number of occurrences of each retrieved image I. Let $D_{N(I)}$ be the occurrences distance, which is calculated by normalizing $N(\cap Ret(I,Pn))$. Assign the weighting factor $Wp2p = [W_D \; W_N]$ to the feature distance $D_{feature}$ and distance measure according to the number of occurrences $D_{N(I)}$, respectively. The similarity ranking is:

$$Rank = W_{p2p} \cdot [D_{feature} \; D_{N(I)}]^T \tag{23}$$

Distributed CBR with Automatic Relevance Feedback

While automatic relevance feedback (ARF) reduces the need for human interaction for relevance feedback [33], integrating ARF to the proposed distributed CBR framework introduces new challenges for repeated requests to multiple peers, which consumes bandwidth and computational resources. To address this issue, an incremental searching mechanism is proposed to reduce the level of transactions between the peers.

As shown in Figure 7(a), peer A originates a query request to its nearest neighbor peer B. Peer B performs the query, and returns the top matched

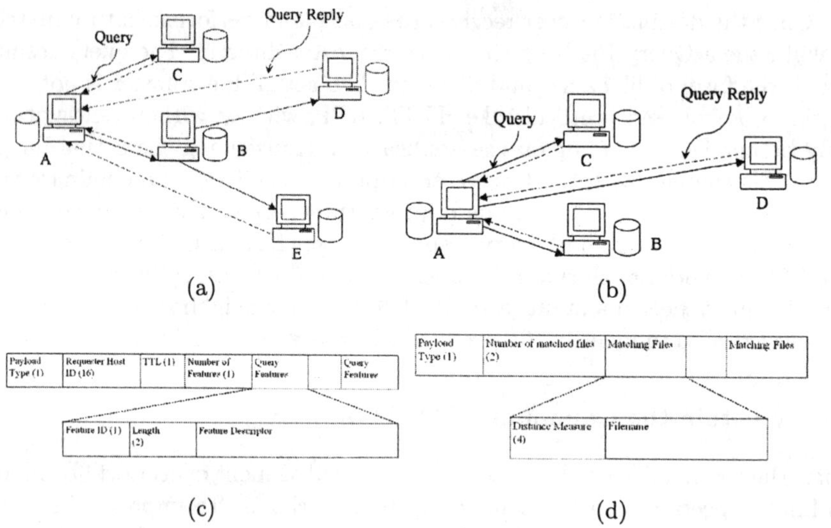

Fig. 6. (a) Neighborhood discovery, (b) Search within community neighborhood, (c) Packet format for Query, (d) Packet format for Query Reply

feature descriptors to peer A. Consequently, Peer A evaluates the retrieval results using the self-organizing tree map (SOTM) algorithm, and generates a new feature vector using RBF method [33]. A new query request using the new feature vector will be sent to peer B, as well as incrementing the audience to peer C. The query request and automated retrieval evaluation process is repeated until a pre-defined number of query peers is reached.

Offline Feature Calculation

As described previously, online feature calculation requires high computational resources and results in delay for content retrieval. Redundant online feature computation can be eliminated by the following specifications:

- Each image stored in the P2P CBR network is attached with its feature descriptor.
- When a peer creates a new image, the feature descriptors will be computed and attached with the image file before announcing the availability of the new image.
- Any image transmission over the distributed P2P network will be coupled with the transmission of the image's feature descriptor.

Advanced Feature Calculation

For extensibility, apart from offline feature calculation, an ideal CBR system should also allow new features to be computed on the fly. These new features,

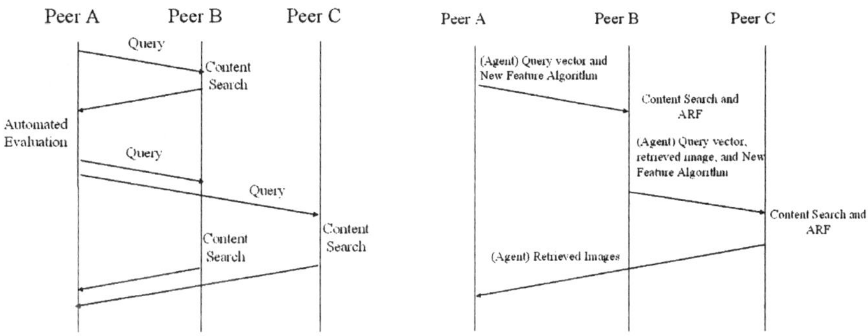

(a) Incremental P2P CBR system (b) Agent-based ARF

Fig. 7. Advance Feature Calculation.

typically address the specific feature for a query image, are best implemented with the ARF such that no additional human interaction is required. To realize advanced feature descriptors for the distributed CBR systems, we propose two approaches: Query Node ARF and Agent-based ARF:

Query Node ARF: As illustrated in Figure 7(a), the query node, Peer A, makes the initial request to the destination node, Peer B. The destination nodes will transfer the retrieved images to the query node. The query node will calculate the advanced feature descriptor on the fly and perform ARF. A new query vector will be generated and a new query will be made from the query node to the destination nodes repeatedly, until a pre-defined number of iterations is met.

Agent-based ARF: The major drawback of the Query Node ARF approach is the bandwidth cost for multiple retrieved image transmission to the query node, as well as the computation cost for the query node to calculate advanced features on-the-fly. We propose an infrastructure applying the software agent technique [20] to offload the bandwidth and computation cost from the query node. As shown in Figure 7(b), the query node, Peer A, initiates a software agent to carry the query vector using standard a feature descriptor, as well as the advanced feature algorithm, to the destination node, Peer B. Peer B performs the retrieval with ARF with the query vector together with the advanced features computed for all the images on-the-fly. The software agent carries the query vector, advanced feature algorithm, as well as the retrieved images from Peer B, to the subsequent neighbor node, Peer C. Upon reaching a pre-defined number of neighbors, the software agent will carry the retrieved images back to the query node.

Offloading computational cost from the query node to destination nodes raises security concern, as the flexibility of remote procedure execution opens the doorway for various malicious attacks. Authenticating as well as validating the integrity of the software agent is therefore a must.

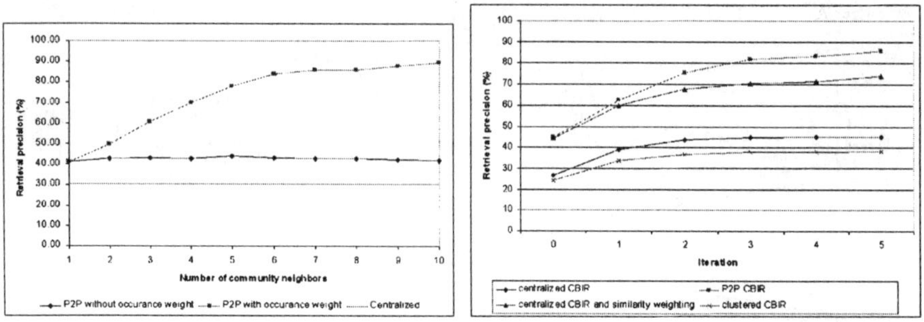

(a) (b)

Fig. 8. (a) Statistical retrieval result for the proposed P2P CBR system, (b) CBR with automatic relevance feedback using RBF and SOTM methods.

4.3 Experimental Results

For experimental purpose, a P2P network is constructed using an evenly distributed tree structure and each peer is connected to five other peers. The number of image categories each peer possesses follows normal distribution, with the mean μ_{cat}=10 and the standard deviation σ_{cat}=2. The number of image per category is also normally distributed, with the mean μ_{images} =50, and the standard deviation σ_{image}=5.

The simulation is performed with the Corel photo image database, which consists of 40,000 color images. The statistical results are taken from averaging the 100 queries in the categories of bird, canyon, dog, old-style airplane, airplane models, fighter jet, tennis, Boeing airplane, bonsai, and balloon.

Figure 8(a) shows the statistical analysis of the size of community neighborhood to the retrieval precision. We observe a steady increase for the retrieval precision against the size of the community neighborhood. Such characteristics serve the foundation of the proposed P2P CBR system.

The same experiment is repeated for clustered CBR systems, where the image database is pre-classified into 10 clusters using the k-means algorithm. Prior to the similarity matching process, the best cluster to perform the retrieval is determined from the query feature descriptor and the cluster centroid using the nearest-neighbor algorithm. As shown in Figure 8(b), the proposed clustered CBR system trades off 5.75% retrieval precision in average for offloading the computation with an order $O(1/N)$ where N=10.

The inter-dependence between each individual image database can be used to improve the retrieval precision for a centralized CBR system, using the same algorithm proposed for a distributed CBR system. While the centralized CBR system typically includes a higher order of database, higher diversity is expected. In our simulation, the number of images per category is also normally distributed, with μ_{cat}=20, σ_{cat}=5, μ_{images} =50 and σ_{image}=20.

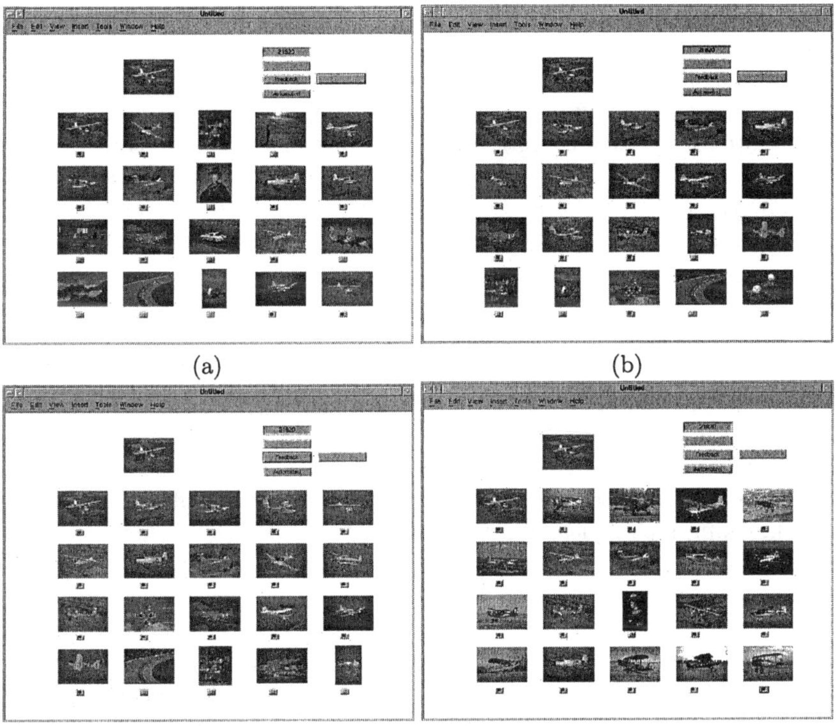

(a) (b) (c) (d)

Fig. 9. (a) CBR from a centralized database, the retrieval precision is 55%, (b) Relevance feedback with single iteration, the overall retrieval precision is 70%, (c) Relevance feedback with five iterations, the overall retrieval precision is 80%, (d) P2P CBR with automated relevance feedback, the overall retrieval precision is 95%

Comparisons between the centralized CBR, the clustered CBR, centralized CBR accounting inter-dependencies between individual databases, and the distributed CBR, are illustrated in Figure 8(b). Accounting the overlap between relevant databases used for distributed P2P CBR, as described in Section 4.2, we observed improvement in the retrieval precision for centralized CBR with similarity weighting.

Finally, screen shots of a query for an airplane, from the centralized database, with the first and fifth iteration of interactive relevance feedback, and with ARF on the distributed P2P CBR system, are shown in Figure 9(a)-(d), respectively.

5 From Content-based to Concept-based Retrieval

Preceding sections have accounted works in content-based retrieval (CBR) from attempts to derive content-based perceptual features, considerations about search strategies in distributed environments, to the use of cues for indexing and retrieval of video data. In this section, we take an elevated post, to review the defining characteristic and usefulness of the current CBR approaches and to articulate any required extension in order to support semantic queries.

5.1 Defining Characteristic of Content-based Retrieval

The defining characteristic of CBR lies with the content-based perceptual features. In many ways, CBR can be identified with modern information retrieval (modern IR). To provide content access to images and audiovisual documents, CBR conveniently drew from modern IR the automation motivation, content-based indexing techniques, and the query-by-example (QBE) paradigm. Accordingly, vector space similarity model, similarity ranking, term-weighting, relevance-feedback, and post-coordinate indexing are not alien concepts in the literature of both practices. Further still, as modern IR relied on relevance feedback to sidestep the difficulty encountered in natural language processing, CBR has made heavy use of relevance feedback to circumvent the problem faced in semantic gap. In addition to these shared characteristics, however, as images and audiovisual documents are perceived by perceptual qualities, CBR has also acquired the capability to derive visual and audio features from audiovisual documents such that these documents can be retrieved by their perceptual characteristics i.e. to query images for visual characteristics, to query audios for aural characteristics, and to query audiovisual documents for multimodal characteristics. Therefore, it is the perceptual features (and not the content-based indexing or QBE) that distinguish CBR from earlier information retrieval practices.

In practice, a small set of perceptual features has been introduced in the literature of CBR. In [21] for example, nearly two dozen perceptual features were identified to constitute the indices of some forty better known CBR prototypes reported in recent time. These features were used to characterize perceptual senses of color, texture, and shape. They were extensively used to serve two distinct types of search in CBR i.e. *perceptual similarity* and *object relevance*. A perceptual similarity search normally uses QBE to obtain a list of documents judged to be perceptually similar—on overall appearance or feature-wise similarity to the query; while an object relevance search intends to return a list of documents of which a certain object forms part of the document.

To this end, it is probably acceptable to conjecture that *perceptual similarity* search has reached maturity for commercial applications. Consequently, the bulk of recent effort has moved to focus on *object relevance search* or more

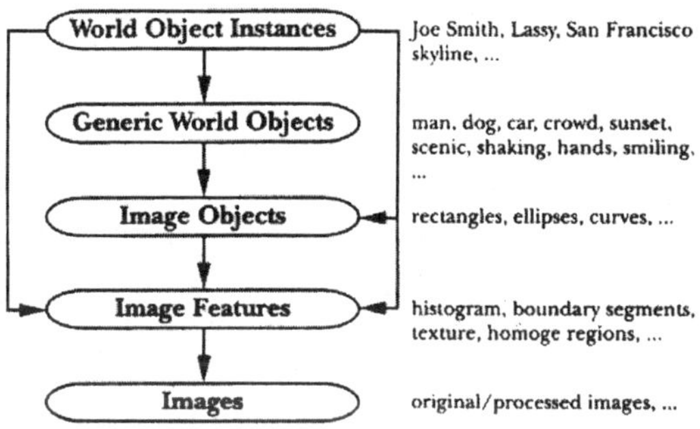

Fig. 10. Semantic abstraction in MUSEUM.

broadly on *semantic retrieval.* The state-of-the-art issue in CBR is thus on addressing *semantic gap*, that is to find ways such that *semantic queries* can be operated by using perceptual features.

5.2 Modalities Come to the Aid of Semantic Search

The quest for objects is a standing challenge. A methodology used to derive objects from perceptual features is depicted in MUSEUM [22] replicated in Figure 10. In this approach, the identification of "Lassy" a pet would be preceded by the spotting of an object class called "dog" that in turn comprises certain geometrical objects associated with certain colors, textures and structural relationships which could be derived from perceptual features. Thus, to operate object-search, content-based image retrieval (CBR) has depended heavily on vision techniques.

Following CBR, earlier works in content-based video retrieval (CBVR) have also attempted to search for objects by visual features. In [23], searches for objects were approached by using local color histograms and color maps. Each of the periodically sampled frames was subdivided into 30x30 cells where local color histogram and bitmaps of the dominant colors were computed and indexed. In a search, corresponding features of the example query were also computed thus allowing the search for content object to be operated by similarity matching of perceptual features.

However, deriving objects from visual features is well-known to be demanding in vision studies. Fortunately, unlike CBIR in which object detection is primarily an image segmentation task, CBVR may take advantage of other perceptual features evident in an audiovisual document. Observably, *salient objects* in a typical video segment are likely to be associated with certain patterns of movement. Thus motion can be used to identify the occurrence of salient objects in a video segment. In [24], moving objects were approximated by motion vectors and the search operation was operated as QBE similarity search.

The fusion of modalities constitutes a significant advance in CBR. It considerably lessens the burden for identifying salient objects in audiovisual documents. Along this line, many heartening works have been reported. In [25], spatiotemporal features were used to compose *salient video stills* by principally mapping moving objects (foreground) of a scene into a high resolution background; in [26], a video sequence was characterized based on a dual hierarchy of spatial and temporal representations; in [27], textual words in video frames were also detected and indexed; in [28], speech recognition was used to detect and index verbal contents of news broadcasts; in [29], *spotting by association* was used to enhance the detection of semantics such that a news segment about *the state of the union speech* could be detected with the help of clues such as visual contexts of *meeting* and *people* as well as the coincidence of *face* and keywords *I*, *believe*, and *think*; and in [30], a probabilistic framework called *multimedia objects* was used to model semantic concepts in the perceptual feature space. Retrieval was then operated as a pattern recognition task over the multimodal features.

Thus far it is clear that the success of a CBR system to support semantic queries is dependent on its capability to derive semantic objects from perceptual features.

5.3 Getting Further with Semantic Retrieval

Ideally the use of low-level perceptual features would afford elegant searches for every semantic concept. Nevertheless, operating semantic retrieval upon perceptual features has encountered numerous problems.

First it evokes high query cost. Traditionally, semantic retrieval was based on QBE-based similarity matching. In light of the incapacity of QBE to support semantic keyword query [31], more recent works have had similarity matching replaced by pattern recognition tasks [12]. For example, rather than to compose an example query, a predefined concept such as *explosion* [30] was detected by using hierarchical HMMs over color histograms, difference between the histograms in the successive frames, and an evenly spaced filter bank for the audio signal. Unfortunately, operating semantic retrieval with pattern recognition tasks also results in high query cost.

Perceptual features based semantic search also entails verbose indices. In contrast to a common belief that perceptual features can facilitate high index exhaustiveness, the operational versatility of perceptual features is nonetheless limited. As similarity in CBR is determined by perceptual features used, semantic retrieval may require a large feature set. For example if *finger print* or *mammogram* documents are to be queried for content concepts; color, texture, shape, and layout features that work for photographs will likely be unbefitting. Likewise, to search for similar faces, a feature like *eigenface* will need to be used. Thus the going is rough; operating a decent semantic retrieval system may require a multitude of perceptual features.

Furthermore, non-verbally expressed information in images and audiovisual documents encompasses concrete and abstract objects. Any attempts to resolve semantic gaps necessarily assume a consistent correlation exists between semantic concepts and perceptual features. Notably top level objects in Figure 10 include a man named Joe Smith, a pet called Lassy, and the scene of the San Francisco skyline. They are concrete objects that have representations in perceptual senses. On the other hand, there are abstract concepts such as *cost*, *information*, and *time* that may not have perceptual representation. These concepts can not be supported naturally by perceptual features.

Consequently, CBR is experiencing significant slow down on its pace of progress. Works on semantic retrieval have been able to entertain only a small set of concepts on a rather small set of documents. Any breakthrough is likely only after great advance in computational intelligence.

5.4 Concept-based Retrieval

In light of the problems faced, a foundational redesign may be warranted to advance semantic retrieval for dealing with a larger set of concepts that operates on a larger set of documents.

Suppose one is building a retrieval system for chess game videos. What usefulness shall this system facilitate? Naturally a chess game video incorporates more than just objects and verbal information. The usefulness thus should not be confined to accessing information expressed in natural languages alone or just to allocate objects such as *pawn*, *rook*, and *queen*; it should also embrace non-verbally expressed game concepts—*check*, *fork*, *king's gambit, positional game*, etc. The challenge thus lies on how non-verbally expressed concepts may be indexed and queried?

To deal with non-verbally expressed information in images and audiovisual documents, a document is better treated as a piece of information bearing medium, having on it the representation of thoughts expressed in certain *languages*. The existence of such languages is necessary to guarantee communications. In other words, expressions in a document will need to be made in accordance with a convention that is understood by its readers; otherwise no communication can be justified. More specifically, a language is meant here to cover all structured and conventional methods of communication including the system of signs and rules. It thus comprises natural languages and other non-verbal systems such as body language, painterly language, music language, and also the signs and rules of the chess game.

Treating documents as the embodiment medium for expressed thoughts of certain languages not only affords a convenient view as to what constitutes the semantic contents of a document, but more importantly the insight into how semantic retrieval can be operated. As a thought is communicated with a certain concept language in a document, the key to concept-based retrieval

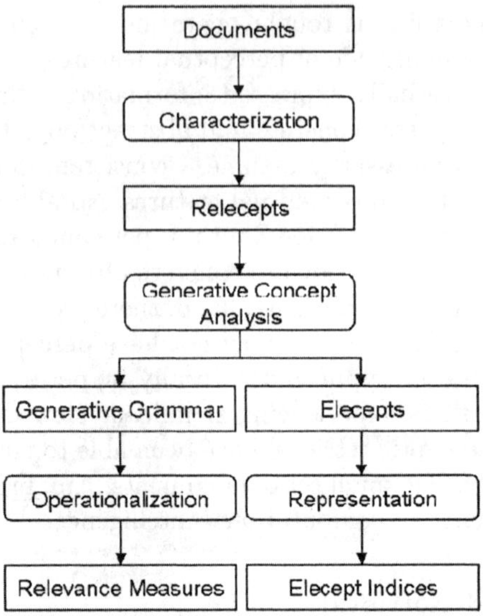

Fig. 11. Basic processes of Concept-based retrieval.

is to allocate the lexicon and grammar of that concept language and to built index and query structures upon them. Perhaps one may draw a convenient analogy to the working of full-text indexing. In the parlance of the latter, concept-based retrieval shall seek to allocate all distinct words of a concept language and the mechanism by which a *more specific concept* is expressed by using words in that language. Subsequently the words are used to index the document collection while the concatenation mechanisms such as *concurrency* (AND) and *adjacency* (ADJ) operators are used to support query operation. Arbitrary concept query may then be supported by post-coordination of words with the concatenation operators. In the vernacular of [32], the basic design of concept-based retrieval is to derive elemental concepts (elecepts) and generative grammar of the concept language and to build index and query structures based upon them.

Figure 11 illustrates the basic processes of concept-based retrieval. First the salient aspects of relevance called relecepts of a database are identified. Each relecept is then subject to a generative concept analysis where elecepts and generative grammar of the concept language are derived. Documents are then indexed with elecept indices, whereas generative grammar is used to operate the query operation.

Elecept indices may extend several advantages. First they support economical indexing. As elecepts are finite discrete entities, a concise description scheme can be devised. For example, to provide semantic search for game concepts in chess game, chess elecepts can be indexed by using shorthand notation instead of traditional perceptual features.

Elecept indices also extend less-demanding burdens on concept detection. As elecepts are finite discrete granular concepts, the operations needed to detect elecepts from a document are likely to comprise less demanding tasks. Sensibly to support the game concept, only movement information is necessary; thus we need to process only those frames in a short moment after a move has taken place. Concept detection thus can enjoy the support of context hints to avoid the need to track the video continuously and expensively.

Furthermore the use of elecept indices also emphasizes a clear distinction between conceptual entity and its representation. Continuing with the chess-game example, clearly elecept indices may be derived from a chess dictionary by using optical-character-recognition (OCR), from existing digital archives where a simple filtering computation is sufficient, or else from a live game video where more demanding vision-based mechanisms are required.

Meanwhile for the query operation, elecept indices can be treated as a vector of elecept terms where numerous query techniques developed in modern IR and CBR can be utilized. Nevertheless, unlike perceptual features, an inverted index of elecepts can also be judiciously devised. As elecepts are essentially vocabulary of a concept language, indices in concept-based retrieval comprise representation of granular concepts rather than simply content-based features. For example, elecept indices for the abovementioned chess-game may encapsulate the information about where the pieces were after each move in the game. Accordingly an inverted index of elecepts can be used to find documents where certain movement concepts occur in chess games. The use of inverted index of elecepts in turn confers an attractive inducement, not only in terms of operational efficiency, but also the extension of queryability as generative grammar operators such as AND, OR, NOT, ADJ, NEAR, WITHIN are used to post-coordinate a large number of semantic queries from the fine grain elemental concept indices.

References

1. Figueiredo, M., Jain, A.K. (2000) Unsupervised selection and estimation of finite mixture models. Proc. Int. Conf. on Pattern Recognition, 87-90.
2. Bilmes J. (1998) A gentle tutorial on the EM algorithm and its application to parameter estimation for Gaussian mixture and hidden Markov models. Technical Report ICSI-TR-97-021, University of Berkeley.
3. Muneesawnag P., Guan, L. (2004) An interactive approach for CBR using a network of radial basis functions. to appear in IEEE Trans. on Multimedia.
4. Broomhead, D. S., Lowe, D. (1988) Multivariable functional interpolation and adaptive networks. Complex Syst., **2**, 321-355.
5. Chen, S., Grant, P. M., Cowan, C. F. N. (1992) Orthogonal least squares algorithm for training multi-output radial basis function networks. IEE Proc. Part F, **139**, 378-384.

6. Corel Gallery Magic 65000 (1999) www.corel.com.
7. Chang, Y.-L., Zeng, W., Kamel, I., Alonso, R. (1996) Integrated image and speech analysis for content-based video indexing. Proc. of IEEE Int. Conf. on Multimedia Computing and Systems, 306-313.
8. Dahyot, R., Kokaram, A., Rea, N., Denman, H. (2003) Joint audio visual retrieval for tennis broadcasts. Proc. of IEEE Int. Conf. on Acoustics, Speech, and Signal Processing, **3**, 561-564.
9. Saraceno, C., (1999) Video content extraction and representation using a joint audio and video processing. Proc. of IEEE Int. Conf. on Acoustics, Speech, and Signal Processing. **6**, 3033-3036.
10. Huang, J., Liu, Z., wang, Y., Chen, Y., Wong, E. K. (1999) Integration of multimodal features for video scene classification based on HMM. IEEE Workshop on Multimedia Signal Processing. 53-58.
11. Jasinschi, R. S., Dimitrova, N., McGee, T., Agnihotri, L., Zimmerman, J., Li, D., Louie, J. (2002) A probabilistic layered framework for integrating multimedia content and context information. Proc. of IEEE Int. Conf. on Acoustics, Speech, and Signal Processing. **2**, 2057-2060.
12. Naphade, M. R., Huang, T. S. (2002) Extracting semantics from audiovisual content: The final frontier in multimedia retrieval. IEEE Trans. on Neural Networks. **13**, 793-810.
13. Muneesawang, P., Guan, L. (2003) Video retrieval using an adaptive video indexing technique and automatic relevance feedback. IEEE Workshop on Multimedia Signal Processing. 220-223.
14. Smith, J. R., Basu, S., Lin, C.-Y., Naphade, M., Tseng, B. (2002) Interactive content-based retrieval of video. Proc. IEEE Int. Conf. on Image Processing. 976-979.
15. Kohonen, T. (1997) Self-organising MAPS, 2nd ed. Springer-Verlag, Berlin.
16. Flickner, M. (1995) Query by image and video content: The QBIC system. IEEE Computer. **28**, 23-32.
17. Smith, J. R., Chang, S. F. (1996) VisualSeek, A fully Automated Content-based Image Query System. Proc. of ACM Multimedia Conference. 87-98.
18. Ng, C. H., Sia, K. C. (2002) Peer Clustering and Firework Query Model. Proc. of the 11th World Wide Web Conference.
19. Lay, J., Guan, L. (2003) SOLO: an MPEG-7 optimum search tool. IEEE International Conference on Multimedia and Expo.
20. Nwana, H. S. (1996) Software Agents: An Overview. Knowledge Engineering Review. **11**, 1-40.
21. Veltkamp, R. C., Tanase, M., Sent, D. (2001) Features in Content-based Image Retrieval Systems: A Survey. A chapter in State-of-the-art in Content-Based Image and Video Retrieval. edited by Veltkamp, R. C., Burkhardt, H., Kriegel, H.-P. Kluwer. 97-124.
22. Mehrotra, R. (1996) Content-based Image Modeling and Retrieval. Proc. of the Clinic on Library Applications of Data Processing. 57-67.
23. Nagasaka, A., Tanaka, Y. (1991) Automatic Video Indexing and Full-Video Search for Object Appearances. Proc. IFIP TC2/WG2.6 Second Working Conf. on Visual Database Systems. 113-127.
24. Ioka M., Kurokawa, M. (1992) Method for Retrieving Sequences of Images on the Basis of Motion Analysis. Proc. of SPIE Vol. 1662: Image Storage and Retrieval Systems. 35-46.

25. Teodosio L., Bender, W. (1993) Salient Video Stills: Content and Context Preserved. Proc. of the first ACM international conference on Multimedia. 39-46.
26. Dimitrova, N., Golshani, R. (1995) Motion Recovery for Video Content Classification. ACM Transactions on Information Systems. **13**, 408-439.
27. Lienhart, R. (1991) Automatic Text Recognition for Video Indexing. Proc. of the fourth ACM International Conference on Multimedia. 11-20.
28. Hauptmann, A. G. (1995) Speech Recognition in the Informedia Digital Video Library: Uses and Limitations. Proc. of the Seventh International Conference on Tools with Artificial Intelligence. 288-294.
29. Nakamura, Y., Kanade, T. (1997) Semantic Analysis for Video Contents Extraction—Spotting by Association in News Video. Proc. of the fifth ACM International Conference on Multimedia. 393-401.
30. Naphade, M. R., Kristjansson, T., Frey, R., Huang, R. S. (1998) Probabilistic Multimedia Objects (Multijects): A Novel Approach to Video Indexing and Retrieval in Multimedia Systems. Proc. IEEE Int. Conf. on Image Processing. **3**, 536-540.
31. Chang, S. F., Smith, J. R., Beigi, M., Benitez, A. (1997) Visual Information Retrieval from Large Distributed Online Repositories. Communications of the ACM. **40**, 63-71.
32. Lay, J. A., Guan, L. (2004) Retrieval for Color Artistry Concepts. IEEE Transactions on Image Processing. **13**, 326-339.
33. Muneesawang, P., Guan, L. (2002) Automatic machine interactions for content-based image retrieval using a self-organizing tree map architecture. IEEE Trans. on Neural Networks. **13**, 821-834.

Vector Color Image Indexing and Retrieval within A Small-World Framework

D. Androutsos,[1] P. Androutsos,[2] K.N. Plataniotis,[2] A.N. Venetsanopoulos[2]

[1] **Ryerson University**
Department of Electrical & Computer Engineering
350 Victoria St.
Toronto, ON, M5B 2K3 CANADA
`dimitri@ee.ryerson.ca`

[2] **University of Toronto**
Edward S. Rogers Sr. Department of Electrical & Computer Engineering
10 King's College Road
Toronto, ON, M5S 3G4 CANADA
`{oracle,kostas,anv}@dsp.toronto.edu`

Abstract. In this chapter, we present a novel and robust scheme for extracting, indexing and retrieving color image data. We use color segmentation to extract regions of prominent and perceptually relevant color and use representative vectors from these extracted regions in the image indices. Our similarity measure for retrieval is based on the angular distance between query color vectors and the indexed representative vectors. Furthermore, we extend *small world* theory and present an alternative approach to centralized image indices using a distributed rationale where images are not restricted to reside locally but can be located anywhere on a network.

Keywords: image retrieval, color, perceptual vector measure, small world theory, distributed indexing

1 Introduction

Content-Based Image Retrieval (CBIR) is a research area dedicated to the image database problem [1, 2]. The number of image database systems which have recently been developed, and others that are currently under development [3–9], is strong evidence to this area's importance. A key aspect of image databases is the creation of robust and efficient indices, which are used for retrieval of image information. In particular, color remains the most important low-level feature, which is used to build indices for database images. Specifically, the color histogram remains the most popular index, due primarily to its simplicity [10, 11].

However, using the color histogram for indexing has a number of drawbacks:

- histograms require quantization to reduce the dimensionality. A typical 24-bit color image generates a histogram with 2^{24} bins, which requires at least 2 megabytes of storage space, depending on the resolution. However, with quantization comes loss of color information and there is no set rule as to how much quantization should be done.
- the color space which is being histogrammed can have a profound effect on the retrieval results and also governs the type and amount of quantization. For instance, for the RGB space, uniform quantization can be used since the distribution of each color plane is essentially uniform. However, in spaces such as the Munsell and LUV, uniform quantization may not suffice [10].
- color *exclusion* is difficult using histogram techniques. For example, a simple way that exclusion could be achieved would be to simply omit histograms which contain non-empty bins corresponding to the exclusion color. Unfortunately, it would be required to specify a-priori *which* bins represent the exclusion colors. If the degree of quantization is also factored in, it is easy to see that determination of these bins is an ill-posed problem. If too *few* bins are selected, then a perceptually similar bin may be included in the similarity calculation and thus not excluded. If too *many* bins are specified, then a bin corresponding to a dissimilar color may be excluded, which could possibly misclassify an otherwise valid image. A more desirable approach would be to allow certain colors to be excluded from a user-defined query right from the start, without requiring an additional level of analysis. In addition, a similarity measure should be used to determine if indexed colors match an exclusion color, and their level of similarity should affect the overall image similarity calculation accordingly.
- the histogram captures *global* color activity; no spatial information is available. To include spatial information requires each image to be partitioned into n regions and a histogram built for each region [12], which consequently requires n times more storage.

We present a novel and robust scheme for extracting, indexing and retrieving color image data[3]. We use color segmentation to extract regions of prominent and perceptually relevant color and use representative vectors from these extracted regions in the image indices. We do away with histogram indexing techniques and instead implement vector techniques. This way we end up with a much smaller index which does not have the over-completeness or *granularity* of a color histogram, yet performs better and robustly. Our sim-

[3]Some sections of this chapter are reprinted from *Computer Vision and Image Understanding*, **Vol 75**(1-2), D. Androutsos et al, "A novel vector-based approach to color image retrieval using a vector angular-based distance measure", Pages 46-58, 1999, with permission from Elsevier.

ilarity measure for retrieval is based on the angular distance between query color vectors and the indexed representative vectors.

Furthermore, we acknowledge the fact that image database need not be locally centralized and address the concept of having image databases distributed over many local and remote sites. Some systems [13, 14] have addressed both distributed and peer-to-peer paradigms [15, 16], but the convenience of centralized techniques for indexing image descriptions is still almost always assumed. To address this, we present an alternative approach to centralized image indices using a distributed rationale where images are not restricted to reside locally but can be located anywhere on a network. This poses problems, especially when nodes of a network (i.e, one or more images) are inaccessible. Unlike centralized approaches, distributed techniques exhibit degradation and increased reliability, in the face of certain types of faults, at the cost of increased storage and complexity.

The outline of this chapter is as follows: Section 2 discusses the feature extraction method and how image indices are built. Section 3 discusses the rationale and model used for our decentralized image database, the *Small World* distributed index (SWIM), and presents the SWIM system and how the database population promotes network growth. Section 4 introduces the vector approach to retrieval and the vector angular-based measure which is implemented, within the SWIM framework, to perform similarity calculation and retrieval. Section 5 introduces and discusses what we call the *Multidimensional Query Distance Space* and how we implement it for multiple color query and color exclusion. Finally, in Section 6 conclusions are drawn.

2 Feature Extraction & Indexing

Recently, some image retrieval systems have begun to move away from histogram techniques and begun to make use of segmentation to extract and index features [17–21]. Color image segmentation is an area which has received a lot of attention and research. Its popularity and effectiveness lies in the fact that the task of color segmentation is an inherent component of human visual processing. At the earliest stages of human vision, low-level processing naturally and automatically partitions a perceived scene, without any recourse to information regarding content or context. These extracted color features are then used in later stages of human vision to build objects which are then identified and classified by the brain.

To extract color features and build indices into our image database we take into consideration factors such as human color perception and recall. For example, as humans it is very difficult, if not impossible, for us to visually discern the difference between two very close RGB values, e.g., $(255, 48, 32)$ and $(254, 48, 32)$. Furthermore, if we were to describe the color content of an image, we would use terms such as *red*, *dark yellow* or *bright green*, not RGB values. In addition, we tend to focus on and remember bright saturated regions and large color regions present in an image. Film manufacturers exploit the former of these characteristic by increasing the saturation of their color film to make colors appear more vivid to better mimic our recall of a photographed scene [22].

When we wish to describe an image or to find a desired image, we essentially build a low-level model of the image in question in our mind and compare candidate images to this model. The color *granularity* provided by histogram indexing is, in most cases, not necessary, especially when the final observer is a human. Thus, it is natural to segment an image into regions of perceptually prominent color and retrieve candidate images based on the similarity to the color of each of these regions.

2.1 Segmentation

Our method of color indexing implements *recursive* HSV-space segmentation to extract regions within the image which contain perceptually similar color. We chose this space, due to its proven segmentation performance and for the fact that it allows for fast and efficient automation. It is not dependent on variables, such as seed pixels or number of extracted colors, as in clustering and region growing techniques [23, 24], and this is of great significance if database population is to be independent of human intervention.

The HSV space [25] classifies similar colors under similar hue orientations. *Hue* is particularly important, since it represents color in a manner which is proven to imitate human color recognition and recall. The conversion from RGB to HSV is performed with the following equations:

$$H_1 = cos^{-1}\left(\frac{\frac{1}{2}[(R-G)+(R-B)]}{\sqrt{(R-G)^2+(R-B)(G-B)}}\right), \tag{1}$$

$$\text{where } H = H_1 \text{ if } B \leq G, \text{ otherwise } H = 360^o - H_1;$$

$$S = \frac{max(R,G,B) - min(R,G,B)}{max(R,G,B)}; \tag{2}$$

$$V = \frac{max(R,G,B)}{255}, \tag{3}$$

where R, G and B are the red, green and blue component values which exist in the range $[0, 255]$.

Here, the peaks of the *hue* histogram, which is known to contain most of the color information, are thresholded, while also taking into account *saturation* and *value* information.

The first step is to build a *hue* histogram for all the *bright chromatic* pixels. We have found experimentally that these tend to be colors that have *value* $>$ 75% and *saturation* $\geq$ 20%. Once the pixels which satisfy this criterion are identified, the *hue* histogram is built and thresholded into m bright colors, where m is an image-dependent quantity determined by the number of peaks which the *h*ue histogram exhibits.

From the remaining image pixels, *saturation* and *value* are used to determine which regions of the image are achromatic. Specifically, it has been found, in the literature and experimentally [26, 27], that colors with *value*$<25\%$ can be classified as **black**, i.e., at the bottom of the HSV cone, and that colors with *saturation*$<20\%$ and *value*$>75\%$ can be classified as **white** as shown in Figure 1.

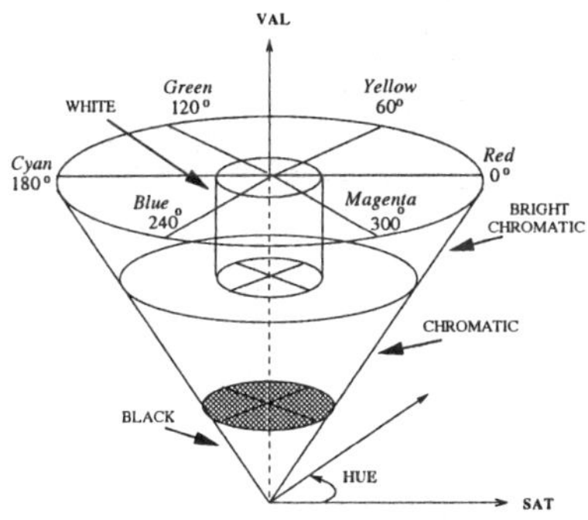

Fig. 1. HSV cone depicting BLACK, WHITE, BRIGHT CHROMATIC and CHROMATIC regions. Reprinted from [41] with permission from Elsevier.

All remaining pixels fall in the *chromatic* region of the HSV cone. However, there may be a wide range of *saturation* values. To account for this, we calculate the *saturation* histogram of all these remaining chromatic pixels. The *saturation* histogram is, in general, multi-modal and we take this fact into account. Many segmentation researchers have classified the *saturation* histogram to be bimodal, however this is not true and it has been found that more accurate color segmentation can be obtained by taking into account its multi-modal nature [27]. Assume that a *saturation* histogram exhibits p peaks, we can threshold each of these peaks and calculate the *hue* histogram for the pixels contained in each given peak. Each resulting *hue* histogram, which exhibits n peaks, is thresholded accordingly to obtain n colors. The process is then repeated for each of the p *saturation* peaks. The entire process is shown in the flowchart in Figure 2.

Thus for each image, the segmentation algorithm extracts c colors, where $c = \sum_p n_p + m$, which is clearly an image dependent quantity. Finally, we calculate the average color of each of the c colors and use that RGB value as each region's representative vector. The reason for using RGB vectors is primarily due to the fact that there is no established method for similarity calculation in the HSV space. In addition, as will be seen in the next section, by using RGB vectors we are able to exploit powerful angular-based distance measures for similarity calculation.

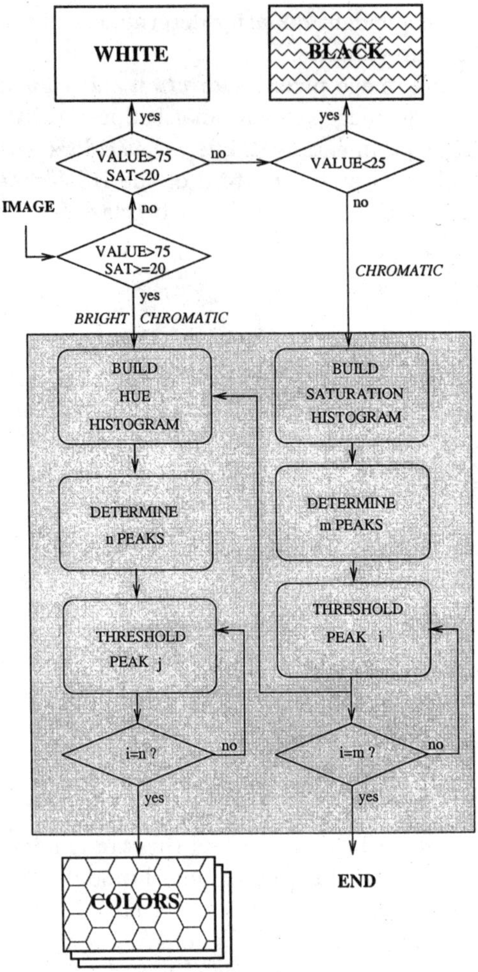

Fig. 2. Flowchart of segmentation procedure. Each image is examined to classify the pixels into one of the four categories: BLACK, WHITE, BRIGHT CHROMATIC and CHROMATIC regions. For the latter two, *hue* and *saturation* histograms are built and the corresponding peaks thresholded to segment the colors. Reprinted from [41] with permission from Elsevier.

Figure 3 shows a typical image, *jellybean*, and the segmented result after our recursive segmentation procedure, where 12 colors were identified and extracted, and each region was filled with its average color. As can be seen, the segmented result provides an accurate low-level representation of the color content of the original image with all dominant and bright colors correctly identified.

The above segmentation technique was performed on our test database of 2400 24-bit images of 512×512 resolution of general image content including

Fig. 3. True-color 24-bit *jellybean* image and its segmented image of only 12 colors. Reprinted from [41] with permission from Elsevier.

natural scenes, people, artwork, architecture, animals and plants. Each image index was built using the c extracted color vectors, along with the percentage of each extracted color present in the image and the number of regions which contain each color. A statistical analysis of our entire 2400 image database revealed that the average number of colors extracted was 4.68, including black and white, and the maximum and minimum number of extracted colors were 13 and 1. This is a surprisingly small number of colors however, as will be seen, retrieval proves to be very effective.

It is also important to note, that by virtue of using color segmentation, we can also incorporate spatial color information into our indices quite easily and efficiently, without having to dramatically increase the index size. However, we do not address spatial color retrieval in this chapter.

3 SMALL WORLD DISTRIBUTED INDEXING

The phrase "*it's a small world*" has proven quite truthful and nearly everyone is connected via an intricate series of complex relationships. Early studies on such '*small world*' acquaintance networks were performed by Pool and Kochen [28] (among others), yet it was Milgram who first put social network theory into action and arrived at the conclusion that the typical number of hops or *degrees of separation* between any two people was approximately equal to six [29]. More recently, small world theory has received renewed interest as a result of the work of Watts and Strogatz in their analysis of networks exhibiting the small world qualities of high *clustering* and low characteristic path length [30]. Similar thinking can be applied to image databases and in particular to extracted image indices. The main concept is that within an image database, all images are related to some degree and each image can keep record of which other images it is most similar to, given selected similarity criteria, (i.e., feature(s) and measure(s)).

3.1 SWIM

A small world paradigm is employed here to create a distributed database of our extracted feature data[4], (i.e., color), through the local storing of peer node index data. The *Small World Image Miner* (SWIM) creates a directed small world network of images which are connected according to color index similarities. This rationale is similar to the way in which World Wide Web pages exist on the Internet, where directed links typically point to additional pages with similar subject matter. Here, each image index in the database contains links to the other images for which it is most similar to. In addition, the SWIM system attempts to mimic the way humans perform referrals and introductions between acquaintances. It is only because humans retain descriptions about the members of their acquaintances that they are able to make introductions between people who they believe are somehow similar (i.e. a decision is made on whether two acquaintances share common interests, backgrounds, etc.).

Under this small world concept, images must become active elements within the index. This is achieved when an image is initially introduced into the database. Once feature extraction is performed to create an index, an additional step is performed by the SWIM agent to store metadata which identifies the most similar peers. This stored data, along with the index data creates a *node* which we call a *SWIMage*[5].

Furthermore, the SWIM agent constantly interacts with new images and search agents by performing distance calculations and acting as a referrer to other nodes. This system is different than other distributed approaches such as that used by *CHORD* [15], as no hash function is used for distributing and locating data, and searches are performed using a decision based agent according to local node data descriptions. Although similar to the DISCOVIR system in [14], SWIM permits multiple levels of index data. In our simplified case here, each *SWIMage*, contains an index which contains only color information (as discussed in Section 2), and thus only one level of index data. However, numerous feature indices representing different image attributes can be extracted and can exist as part of the *SWIMage*. In addition, the SWIM system locally stores descriptor data at every node, and promotes searching through the exclusive use of directed links. These attributes make this framework very effective and robust.

As shown in Figure 4, each *SWIMage* consists of an agent for providing interaction between other SWIM network elements, multiple indices for tracking self and peer indices, and directed links to similar peer *SWIMages*.

Similar to the Newman-Watts model [31], a SWIM network consists of a regular lattice merged with what seems like a random graph. Each SWIM network layer, L_ℓ, establishes F directed nearest-neighbor links (friends) between vertices using an external distance measure D_ℓ in the creation of its

[4]Interestingly, any type of features can be extracted and used as indices. For example, texture, shape or even MPEG-7 descriptors could be used.

[5]we use the term *node* and *SWIMage* interchangeably

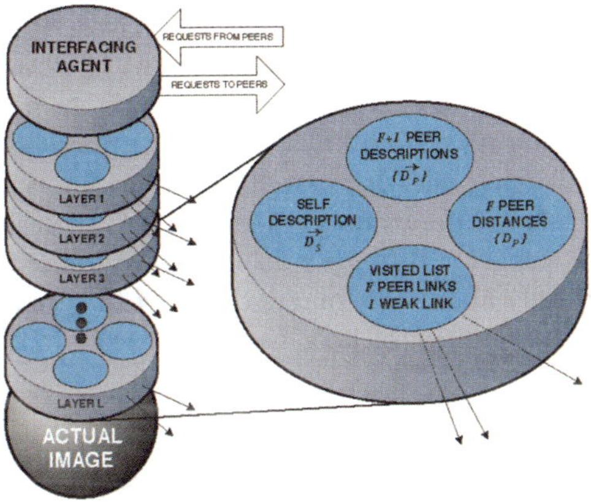

Fig. 4. In addition to actual pixel data, a *SWIMage* consists of a set description layers and an interfacing agent for communication between other nodes. Each description layer stores self and peer description data vectors as well as directed links to similar *SWIMages*.

Similarity Graph, $G_{S\ell}$. A single, directed *weak* link between successive nodes is used to generate the *Weak Graph*, G_W, which is independent of the index or distance measure used. These two components are illustrated in Figures 5(a) and (b).

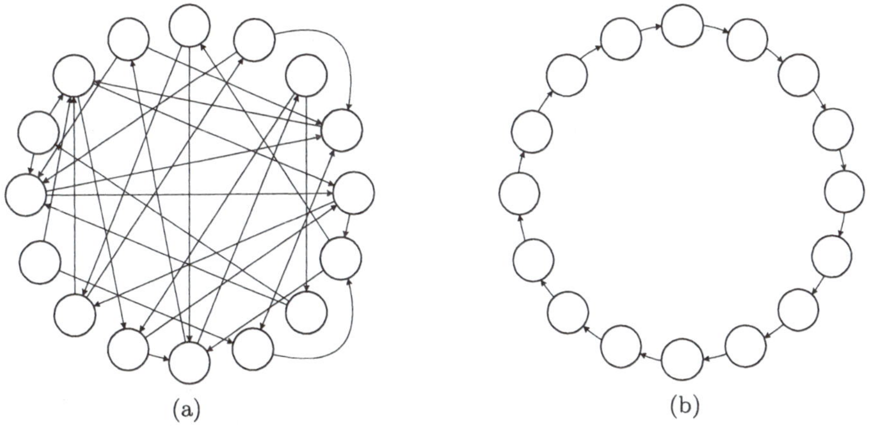

Fig. 5. A 16-node SWIM network. Connections are made (a) from each node to $F = 2$ friends according to an external similarity metric to create $G_{S\ell}$ and (b) between successive nodes to create G_W.

The distance measure employed, (See section 4.2), establishes the nearest neighbors for each node, and must be appropriately chosen with respect to the index of the feature on which the query is based. With this concept in mind, the graph $G_{S\ell}$ can alternatively be thought of as a directed F-lattice within the layer's index space, while G_W represents an incrementally grown graph which (assuming no broken links) serves the purpose of ensuring graph connectivity and allowing network search agents to jump out of tightly connected cliques [32]. It is important to mention here that since peer links are directed in nature, target *SWIMages* do not store index information about their referrers, but only about the images to which they point. This translates to a reduction in the number of peer nodes that each *SWIMage* must keep track of.

3.2 SWIM NETWORK GROWTH

To stay in line with the concept of a distributed indexing scheme, it is imperative that the method used in establishing node connections be as de-centralized as possible. For achieving this, an incremental growth and indexing algorithm is employed by each SWIMage to both ensure connectivity among all SWIM elements, and make certain that each SWIMage has compared itself to all other peers. The assumptions of an existing SWIM network as well as the random introduction (insertion) of new SWIMages to established ones are made. This algorithm consists of two phases; an *introduction phase*, followed by an *exploratory phase*. The introduction phase involves communication between a new SWIMage j and an insertion (or introducing) SWIMage k, in which node j is passed node k's peer information and then incorporated into the weak graph G_W. The exploratory phase begins as nodes j and k perform mutual distance calculations on their respective index vectors $\overrightarrow{D_k}$ and $\overrightarrow{D_j}$ to determine d_{kj}. The algorithm outlined in Figure 6 shows the steps taken by each newly introduced SWIMage. During each node visit, descriptor information is compared according to the given distance measure, and if necessary, exchanged index data is locally stored by each SWIMage. Because only newly introduced SWIMages 'hop' from node to node, there is no need for existing elements to be informed of the existence of new ones; new SWIMages automatically find them by traversing the SWIM network.

4 Retrieval

The small world framework is a distributed approach that allows images to exist at physically separate locations. Search and retrieval time is reduced by the fact that each image index stores information of its closest image matches (wherever they may exist), at the instant that the image is introduced to the database.

The actual retrieval process is performed using a search agent. In a fashion similar to how new SWIMages are introduced into the small world network, search agents are inserted at a starting node, and then proceed to 'hop' to

LOOP (While j target queue != NULL)
SWIMage k passes friend list to j's target queue.
SWIMage k passes weak link w_k to j's target queue.
IF (SWIMage $j \notin G_W$)
SWIMages j and k swap weak links w_j and w_k.
(new SWIMages weakly link to themselves)
Both i and k calculate mutual distance d_{jk} (equal to d_{kj}).
IF ($d_{kj} < d_{k,F}$)
Node k appropriately inserts $f_{new} = (d_{kj}, j, \overrightarrow{D_j})$
into friend list and culls to F friends.
IF (SWIMage j has fewer than F friends)
Node j appropriately inserts $f_{new} = (d_{jk}, k, \overrightarrow{D_k})$
into friend list.
ELSE IF ($d_{jk} < d_{j,F}$)
Node j appropriately inserts $f_{new} = (d_{jk}, k, \overrightarrow{D_k})$
into friend list and culls to F friends.
Node j 'hops' to next node in it's target queue
(nodes to be visited).

Fig. 6. Algorithm for SWIM index growth.

subsequent nodes. Each hop is determined by the distance measure calculation performed between the agent's search criteria and the self and peer index data stored by the currently visited SWIMage. This calculation is done via the *vector approach* discussed next.

4.1 Vector approach

By virtue of the fact that our color indices are 3-dimensional vectors which span the *RGB* space, we have at our disposal a number of vector distance measures that can be implemented for retrieval. However, studies have shown that measures based on the angle of a color vector produce perceptually accurate retrieval results in the RGB domain [33]. Furthermore, angular measures are *chromaticity*-based, which means that they operate primarily on the orientation of the color vector in the RGB space and therefore are more resistant to intensity changes.

As further evidence to the validity of angular-based measures, we find that they exhibit excellent performance in the area of image filtering [34–36]. At first, retrieval and filtering seem unrelated. However, the fact is that both use distance measures to determine candidacy. In particular, Order Statistics filtering implements distance measures to group similar vectors together and discard outliers whereas retrieval *ranks* the similarity between candidates.

Figure 7 depicts the color variation of a given color vector in the RGB space as the angle is varied at 8 points around the central vector. From the figure we can see that as the angle increases further away from the central vector,

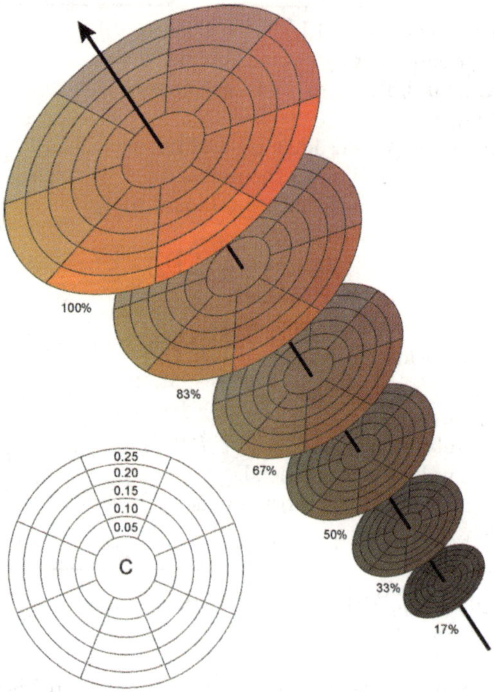

Fig. 7. 6 swatches depicting color change around a central color vector **C**, with *RGB* values of (187, 83, 78), due to changes in the angular distance with respect to that vector. The angle is calculated for angles of 0.05, 0.1, 0.15, 0.20, 0.25 and 0.5 *rad* for 8 points around **C**. Each swatch shows the effect at a different point along **C**, specifically at 125%, 100%, 75%, 50% and 25% of |**C**|. Reprinted from [41] with permission from Elsevier.

the perceived color also changes. However for the small angles of 0.05 *rad* and 0.10 *rad* the color is perceptually the same as the central color, thus a small neighborhood around a given vector in the RGB space contains colors that can be considered equivalent.

4.2 Distance Measure

In our system we implement a combination distance measure composed of an angle and magnitude component [37]:

$$\delta(\mathbf{x}_i, \mathbf{x}_j) = 1 - \underbrace{\left[1 - \frac{2}{\pi} cos^{-1}\left(\frac{\mathbf{x}_i \cdot \mathbf{x}_j}{|\mathbf{x}_i||\mathbf{x}_j|}\right)\right]}_{angle} \underbrace{\left[1 - \frac{|\mathbf{x}_i - \mathbf{x}_j|}{\sqrt{3 \cdot 255^2}}\right]}_{magnitude}, \tag{4}$$

where $\mathbf{x}_i$ and $\mathbf{x}_j$ are 3-dimensional color vectors. Since we deal with RGB vectors, we are constrained to one quadrant of the Cartesian space. Thus, the normalization factor of $\frac{2}{\pi}$ in the *angle* portion is attributed to the fact that the maximum angle which can possibly be attained is $\frac{\pi}{2}$. Also, the $\sqrt{3 \cdot 255^2}$ normalization factor, in the *magnitude* part of (5), is due to the fact that the maximum difference vector which can exist is $(255, 255, 255)$ and its magnitude is $\sqrt{3 \cdot 255^2}$. Both normalization factors contribute so that $\boldsymbol{\delta}$ takes on possible values in the range $[0, 1]$.

This distance measure takes into consideration both the angle between two vectors and the magnitude of the vector difference. However, when two vectors under consideration are collinear, only magnitude difference is used, which is required since two vectors can be collinear but perceptually quite different due to magnitude differences.

For comparison purposes, we have also investigated a number of other common vector distance measures. Specifically, we investigated the *angular distance* between two vectors [34]:

$$\theta(i,j) = 1 - \frac{2}{\pi} cos^{-1}\left(\frac{\mathbf{x}_i \cdot \mathbf{x}_j}{|\mathbf{x}_i||\mathbf{x}_j|}\right), \tag{5}$$

which is the *angle* component of (5) above, the generalized *Minkowski* metric [24](L_M norm), for the three cases $M = 1, 2, \infty$, and the *Canberra* distance metric [38].

4.3 Single Color Query

Our first test of the system deals with single color query, i.e., where only one color is specified. For each image index, the system takes a specified query color and calculates the given distance measure to *each* of the n representative vectors and retains the minimum. The result is the closest match of the set of n representative color vectors, of a given database image, to the query color. The process is repeated for every database index and the entire set of distances are then ranked to retain the top 25 images.

For our results, we chose to look for images which contained at least 50% of the query color *sea green* with *RGB* values (130, 164, 53), which is shown in Figure 8, along with the database image *bat* from which it was extracted. Figure 9 shows the retrieval result using our *combination* distance along with the retrieval result of the other vector distance measures discussed in Section 4.2. In addition, we have also included results using histogram techniques to compare against our technique in Figure 10, specifically we tested RGB and HSV-space histogram techniques. Unavoidably, quantization was performed in the two color spaces. Due to the relatively uniform distribution of the RGB bands, we chose 8 uniform quantization bins for each of the RGB bands. For the HSV color space histogram, we took the statistical nature of the HSV space into consideration, along with human sensitivity to *hue* and *saturation*, and uniformly quantized the *hue*, *saturation* and *value* histograms into 12,5,5 bins respectively [10]. The similarity metric that we implemented for the actual retrieval was the *histogram intersection* [39].

Fig. 8. *bat* image and the color *sea green.* Reprinted from [41] with permission from Elsevier.

Figure 9(a) depicts the top 25 retrieval results for our *combination* measure. Figure 9(b)-(d) show the results for the other discussed vector distance measures. In addition, Figure 10 shows the results for the *RGB* and *HSV*-space histogram retrieval techniques. In each of these figures, the *top left* image is the *sea green* query color and the similarity ranking is from highest to lowest from left to right, top to bottom.

All methods returned images that contained colors similar to the query color however the *combination* measure and the *angle* measure returned more images which were perceptually more accurate. This was established by comparing the retrieved results with a *query set* Q which contains the images which most humans would consider to fit the given query. In our case, The query set Q was obtained from 25 volunteers who were asked to manually search through our 2000 image database and list the images which were considered to contain at least 50% of the query color (*sea green*). The results were then tabulated and the top 25 images which were chosen the most often comprised the query set Q, depicted in Figure 11.

Comparing Figures 9 & 10 with Figure 11, we see that our *combination* distance retrieves more images that belong to Q than the other investigated measures and techniques. Specifically, our measure returned 16 images from the query set Q. This is of utmost benefit and importance since the perceived color content of an image is, in most cases, of greater significance than the actual pixel values. In addition, the *combination* measure returned less images that were erroneous, i.e., that contained colors that were totally irrelevant with respect to the query color.

Quantitative performance was evaluated by calculating the retrieval rate, defined as [12]:

$$R_{i,j} = \frac{N_j}{N_i} \times 100, \tag{6}$$

(a)

(b)

(c)

(d)

(e)

Fig. 9. Retrieval result using (a) *combination* distance, (b) *angular distance*, (c) L_1 norm, (d) L_2 norm and (e) L_∞ norm. Top left image of each result is the query color *sea green*. Reprinted from [41] with permission from Elsevier.

where N_i is the total images in a given query set Q, (i.e., all images in the database which match the query), and N_j are the number of images which appear in the top N_i retrieval positions which are part of Q. Table 1 lists the retrieval rates for the above mentioned distance measures. Clearly, it can be seen that the angular-based measures exhibit much higher retrieval rate, and in particular our *combination* measure provided the highest retrieval rate. It even surpassed the two histogram indexing techniques.

(a)

(b)

Fig. 10. Retrieval result using (a) RGB histograms with (8,8,8) quantization bins and (b) HSV histograms with (12,5,5) quantization bins. Top left image is the query color *sea green*. Reprinted from [41] with permission from Elsevier.

Fig. 11. Query set Q obtained from 25 test subjects who were asked to find images which contained at least 50% of the query color *sea green* (top left image). Reprinted from [41] with permission from Elsevier.

Table 1. Retrieval rate for 6 different vector distance measures and 2 histogram techniques.

measure	$R_{i,j}$
L_1	44
L_2	48
L_∞	36
canberra	40
angle	60
combo	64
$RGB_{8,8,8}$	48
$HSV_{12,5,5}$	28

5 Multidimensional Query Distance Space

During the query process, for each user-specified query color, a distance is calculated using (5), to *each* representative vector in a given database index. For multiple color queries, a distance to each representative vector is calculated for each query color. We take the minimum of these distances and form a *multidimensional query distance vector* **D**:

$$\mathbf{D}(d_1,\ldots,d_n) = (min(\delta(\mathbf{q}_1,\mathbf{i}_1),\ldots\delta(\mathbf{q}_1,\mathbf{i}_c)),\ldots,min(\delta(\mathbf{q}_n,\mathbf{i}_1),\ldots\delta(\mathbf{q}_n,\mathbf{i}_c))), \tag{7}$$

where $\mathbf{q}_m$ are the n query colors and $\mathbf{i}_c$ are the c indexed representative colors of each database image.

For example, let's assume that a query consists of 2 query colors, $\mathbf{q}_1$ and $\mathbf{q}_2$ and a given index contains representative color vectors $\mathbf{i}_1$, $\mathbf{i}_2$ and $\mathbf{i}_3$. The minimum distance d_1 between $\mathbf{q}_1$ and $\mathbf{i}_{1\ldots3}$ is taken, along with the minimum distance d_2 between $\mathbf{q}_2$ and $\mathbf{i}_{1\ldots3}$, to build $\mathbf{D}(d_1,d_2)$.

This set of $\mathbf{D}$ vectors span what we call the *multidimensional query distance space* (MQDS) [40]. For the case of 2 query colors, the space is two-dimensional, for 3 colors three-dimensional, etc. Each database image exists at a point in this space and its location determines its retrieval ranking for the given query. The key to this ranking lies in the origin of the MQDS and the *equidistant line*, where all component values of $\mathbf{D}$ are equal (see Figure 12(a)). The database image that is the closest match to *all* the given query colors $q_1, q_2, \ldots, q_n$ is the one which is closest to the origin of the MQDS. This implies that the distance vector $\mathbf{D}$ that is most centrally located, i.e, is collinear with the *equidistant* line of the MQDS and at the same time has the smallest magnitude, corresponds to the image which contains the best match to *all* the query colors.

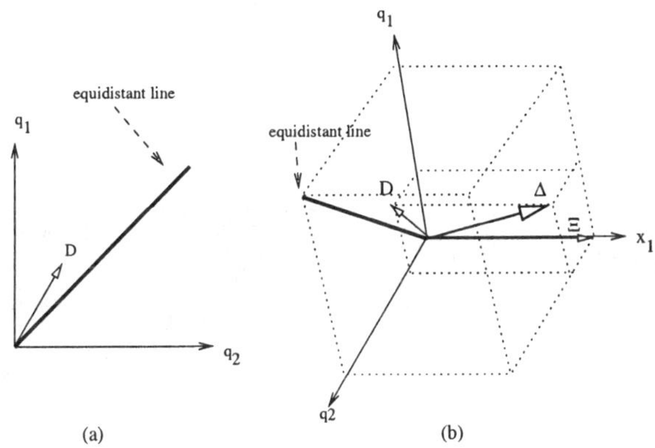

Fig. 12. (a) Vector representation of 2 query colors $q_1 \& q_2$, their multidimensional distance vector $\mathbf{D}$ and the corresponding equidistant line. (b) the same 2 query colors, 1 exclusion color, x_1 and the resulting multidimensional distance vector $\mathbf{\Delta}$. Reprinted from [41] with permission from Elsevier.

Thus, we need to rank the retrieval results based on the magnitude of $\mathbf{D}$ and the angle, $\angle\mathbf{D}$, that it makes with the *equidistant* line. To do this we combine the two values using a weighted sum:

$$\mathcal{R} = w_1|\mathbf{D}| + w_2\angle\mathbf{D}, \tag{8}$$

where lower rank values $\mathcal{R}$ imply images with a closer match to both the query colors. The weights w_1 and w_2 can be adjusted to control which of the two parameters, i.e., magnitude or angle, are to dominate. We have found that values of $w_1 = 0.8$ and $w_2 = 0.2$ give the most robust results. This is to be expected since collinearity with the equidistant line does not necessarily imply a match with *any* query color. It implies that each query color is equally close to the indexed colors. However, as $|\mathbf{D}| \rightarrow 0$, implies closer matches to one or more colors. Thus, a greater emphasis must be placed on the magnitude component.

5.1 Multiple Color Query

Figure 13(a) depicts a user query for two colors. Specifically, it was desired to find images with at least 10% of the RGB colors $26, 153, 33$ (*green*) and $200, 7, 25$ (*red*). Clearly, the displayed top 10 results exhibit colors with strong similarity to the two query colors.

Figure 14 shows a spatial representation of the retrieval results for the same query colors. Here we can clearly see how close the retrieved images are to each query color, and how the vector representation of the two distances determines the retrieved image ranking. Clearly, the origin represents the best match to both colors. Images which lie on the equidistant line have equal distance to each query color however, these images must also be close to the origin to be *similar* matches to the query colors. This reasoning is easily extended to a higher dimension (i.e., a greater number of query colors) with the origin remaining the best retrieval result.

(a) (b)

Fig. 13. Query result for (a) images with *red* & *green*, (b) images with *red* & *green* and excluding *yellow*. Reprinted from [41] with permission from Elsevier.

5.2 Color exclusion

Our proposed vector approach provides a framework which easily accepts exclusion in the query process. It allows for image queries containing any number of colors to be excluded in addition to including colors in the query results. From the discussion in Section 4.1 above, we are interested in distance

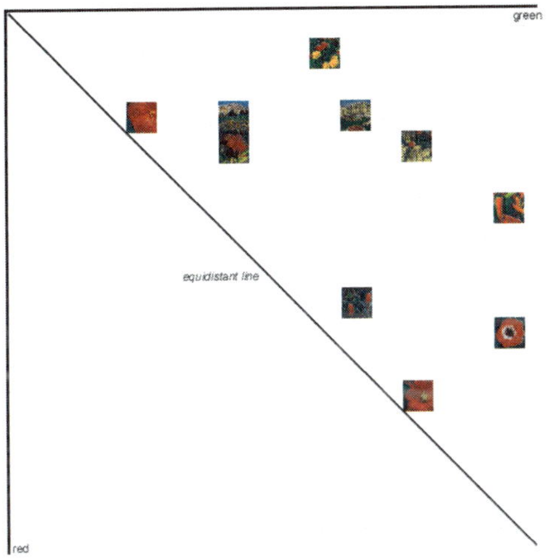

Fig. 14. Spatial representation of the query result for images with *red & green*. The axes represent the calculated distance $\boldsymbol{\delta}$ for each image to the corresponding color. The origin at the top left is the closest match to both query colors. Reprinted from [41] with permission from Elsevier.

vectors **D** which are collinear with the equidistant line and which have small magnitude, i.e., close to the MQS origin. The exclusion of a certain color should affect **D** accordingly and its relation to the equidistant line and the origin. For example, if it is found that an image contains an indexed color which is close to an exclusion color, the distance between the two can be used to either pull or push **D** closer or further to the ideal and accordingly affect the retrieval ranking of the given image.

To this end, we determine the minimum distances of each exclusion color with the indexed representative colors, using (5), to quantify how close the indexed colors are to the exclusion colors:

$$\mathbf{X}(x_1, \ldots, x_n) = (min(\delta(\boldsymbol{\xi}_1, \mathbf{i}_1), \ldots \delta(\boldsymbol{\xi}_1, \mathbf{i}_c)), \ldots, min(\delta(\boldsymbol{\xi}_n, \mathbf{i}_1), \ldots \delta(\boldsymbol{\xi}_n, \mathbf{i}_c))), \quad (9)$$

where $\boldsymbol{\xi}_n$ are the n exclusion colors and $\mathbf{i}_c$ are the c indexed representative colors of each database image. Equation (9) quantifies how *close* any indexed colors are to the exclusion colors. Thus, since x_n depicts similarity, a simple transformation of $1 - x_n$ depicts dissimilarity. Thus, we can apply this transform to each of the components of **X** and then merge this with **D** to give the overall multidimensional vector:

$$\boldsymbol{\Delta} = [\mathbf{D} \quad \mathbf{I} - \mathbf{X}], \quad (10)$$

where **I** is a vector of size n with all entries of value 1. The dimensionality of $\boldsymbol{\Delta}$ is equal to the *# of query colors + # of exclusion colors*.

The final retrieval rankings are then determined from $|\boldsymbol{\Delta}|$ and the angle which $\boldsymbol{\Delta}$ in (10) makes with the equidistant line of the *query* colors. Figure

12(b) graphically depicts the notion of color exclusion and shows how $\mathbf{\Delta}$ affects $\mathbf{D}$ and its relation to the origin and equidistant line.

Figure 13(b) depicts the query result when at least 10% of the *RGB* colors $26, 153, 33$ (green) and $200, 7, 25$ (*red*) ,(same colors as in Section 5.1), were desired and the color $255, 240, 20$ (*yellow*) was excluded. Clearly, images which contained colors close to *yellow* were removed from the top ranking results, as compared to the Figure 13(a) where *yellow* was not excluded.

6 Conclusions

In this chapter we presented a vector-based scheme for color image indexing and retrieval within a distributed small-world framework. We perform a recursive *HSV*-space segmentation technique to identify perceptually prominent color areas and use the average color vector of these areas as image indices into the database. Due to the vector nature of our system, we implement a *combination* distance measure, based on the vector angular distance, in the retrieval process. The query mechanism proves very flexible, allowing single and multiple color query via a *multidimensional query space*. In addition, color *exclusion* can be implemented where certain colors can be specified to *not* be present in the query results. We have tested our system and our distance measure against other common vector distance measures and also against popular histogram indexing schemes and found that it outperforms them all. The retrieval results exhibit the highest retrieval rate of all the measures and methods investigated.

References

1. Tamura H, Yokoya N (1984) Image Database Systems: A Survey. Pattern Recognition 17:1
2. Gudivada VN, Raghavan VV (1995) Content-based image retrieval systems. Computer 28(9)
3. Niblack W, Barber R, Equitz W, Flickner M, Glasman E, Petkovic D, Yanker P, Faloutsos C, Taubin G (1993) The QBIC project: Querying images by content using color, texture and shape. Storage and Retrieval for Image and Video Databases SPIE-1908
4. Gevers T, Smeulder AWM (2000) PicToSeek: Combining Color and Shape Invariant Features for Image Retrieval. IEEE Transactions on Image Processing 9(1)
5. Aggarwal G, Ashwin TV, Ghosal S (2002) An Image Retrieval System With Automatic Query Modification. IEEE Transactions on Multimedia 4(2)
6. Smith JR, Chang S-F (1996) VisualSEEK: A fully automated content-based image query system. ACM Multimedia 96
7. Petland A, Picard RW, Sclaroff S (1994) Photobook: Tools for content-based manipulation of image databases. Storage and Retrieval for Image and Video Databases II SPIE-2185

8. Sethi IK, Coman I, Day B, Jiang F, Li D, Segovia-Juarez J, Wei G, You B (1998) Color-WISE: A system for image similarity retrieval using color. Storage and Retrieval for Image and Video Databases VI, pp 140-149
9. Veltkamp RC, Tanase M (2000) Content-Based Image Retrieval Systems: A Survey. Technical Report, Department of Computing Science, Utrecht University
10. Wan X, Jay Kuo C-C (1995) Color distribution analysis and quantization for image retrieval. Storage and Retrieval for Image and Video Databases IV SPIE-2670:8-16
11. Stricker M, Orengo M (1995) Similarity of color images. Storage and retrieval for image and video databases III SPIE-2420:381-392
12. Zhang H, Gong Y, Smoliar SW (1995) Image retrieval based on color features: An evaluation study. Digital Image Storage and Archiving Systems SPIE-2606, pp 212-220
13. Brunelli R, Mich O (2000) Compass: An image retrieval system for distributed databases. Proc. ICME 2000, pp 145-148
14. Ka Cheung S, et al. (2003) Bridging the P2P and WWW divide with DISCOVIR. Proceedings of the 12th WWW Conference
15. Stoica I, et al. (2001) Chord: A scalable peer-to-peer lookup service for internet applications. ACM SIGCOMM 2001, pp 149-160
16. Clarke I, et al. (2000) Freenet: A distributed anonymous information storage and retrieval system. Lecture Notes in Computer Science, H. Federrath (ed), vol 2009, Springer-Verlag, Berlin
17. Ma WY, Deng Y, Manjunath BS (1997) Tools for texture/color based search of images. Human Vision and Electronic Imaging II SPIE-3016:496-505
18. Bimbo AD, Mugnaini M, Pala P, Turco F, Verzucoli L (1997) Image retrieval by color regions. International Conference on Image Analysis and Processing, pp 180-187
19. Carson C, Belongie S, Greenspan H, Malik J (1997) Region-based image querying. CVPR '97 Workshop on Content-Based Access of Image and Video Libraries
20. Belongie S, Carson C, Greenspan H, Malik J, (1998) Color- and texture-based image segmentation using the expectation-maximization algorithm and its application to content-based image retrieval. ICCV '98
21. Howe NR (1998) "Percentile blobs for image similarity. IEEE Workshop on Content-Based Access of Image and Video Databases
22. Sangwine SJ, Horne REN (eds) (1998) The Colour Image Processing Handbook. Chapman & Hall, 1998.
23. Celenk M (1990) A color clustering technique for image segmentation. Computer Vision, Graphics and Image Processing 52:145-170
24. Duda RO, Hart PE (1973) Pattern Classification and Scene Analysis. John Wiley & Sons, New York
25. Smith AR (1978) Color gamut transform pairs. SIGGRAPH 78, pp 12-19
26. Herodotou N, Plataniotis KN, Venetsanopoulos AN (1998) A content-based storage and retrieval scheme for image and video databases. Visual Communications and Image Processing '98 SPIE-3309
27. Gong Y, M. Sakauchi M (1995) Detection of regions matching specified chromatic features. Computer Vision and Image Understanding 61(2)
28. Pool I, Kochen M (1978) Contacts and influence. Social Networks 1(1):5-51

29. Milgram S (1967) The small world problem. Psychology Today 2:60-67
30. Watts DJ, Strogatz SH (1998) Collective dynamics of 'small-world' networks. Nature 393:440-442
31. Newman MEJ, Watts DJ (1999) Renormalization group analysis of the small-world network model. Physics Letters A 263:341-346
32. Granovetter MS (1973) The strength of weak ties. The American Journal of Sociology 78(6):1360-1380
33. Androutsos D, Plataniotis KN, Venetsanopoulos AN (1998) Distance Measures for Color Image Retrieval. International Conference on Image Processing '98
34. Trahanias PE, A.N. Venetsanopoulos AN (1993) Vector directional filters: A new class of multichannel image processing filters. IEEE Transactions on Image Processing 2(4):528-534
35. Plataniotis KN, Androutsos D, Vinayagamoorthy S, Venetsanopoulos AN (1996) Color Image Processing using Adaptive Multichannel Filters. IEEE Transactions on Image Processing 6(7):933-949
36. Trahanias PE, Karakos D, Venetsanopoulos AN (1996) Directional processing of color images: Theory and experimental results. IEEE Transactions on Image Processing 5(6):868-880
37. Androutsos D, Plataniotis KN, Venetsanopoulos AN (1999) A vector angular distance measure for indexing and retrieval of color. Storage & Retrieval for Image and Video Databases VII SPIE-3656, pp 604-613
38. Johnson RA, Wichern DW (1998) Applied Multivariate Statistical Analysis, Prentice Hall
39. Swain MJ, Ballard DH (1991) Color indexing International Journal of Computer Vision 7(1)
40. Androutsos D, Plataniotis KN, Venetsanopoulos AN (1998) Efficient color image indexing and retrieval using a vector-based scheme. 1998 IEEE Second Workshop on Multimedia Signal Processing
41. Androutsos D, Plataniotis KN, Venetsanopoulos AN (1999) A novel vector-based approach to color image retrieval using a vector angular-based distance measure. Computer Vision and Image Understanding 75(1/2)

A Perceptual Subjectivity Notion in Interactive Content-Based Image Retrieval Systems

Kui Wu and Kim-Hui Yap

School of Electrical and Electronic Engineering, Nanyang Technological University, Nanyang Avenue, Singapore 639798

Abstract. This chapter presents a new framework called fuzzy relevance feedback in interactive content-based image retrieval (CBIR) systems. Conventional binary labeling scheme in relevance feedback requires a hard-decision to be made on the relevance of each retrieved image. This is inflexible as user interpretation varies with respect to different information needs and perceptual subjectivity. In addition, users tend to learn from the retrieval results to further refine their information priority. It is, therefore, inadequate to describe the users' fuzzy perception of image similarity with crisp logic. In view of this, a fuzzy framework is introduced to integrate the users' imprecise interpretation of visual contents into relevance feedback. An efficient learning approach is developed using a fuzzy radial basis function network (FRBFN). The network is constructed based on hierarchical clustering algorithm. The underlying network parameters are optimized by adopting a gradient-descent-based training strategy due to its computational efficiency. Experimental results using a database of 10,000 images demonstrate the effectiveness of the proposed method.

Keywords: content-based image retrieval, fuzzy decision, relevance feedback, user information need

1 Introduction

The rapid growth of multimedia information has increased the demand for efficient access to multimedia data in many applications, ranging from digital image libraries, biomedicine, military to education. Image is an important media which has been used extensively. Content-based image retrieval (CBIR) has been developed as an instrumental approach of accessing image data. It is designed to retrieve a set of desired images from an image collection on the basis of visual contents such as color, texture, shape and spatial relationship that are present in the images. Traditional text-based image retrieval uses keywords to annotate images. This involves significant amount of human labor in manual annotation of large-scale image databases. In view of this, CBIR is proposed as an alternative to text-based image retrieval. Many research and commercial CBIR systems have been developed, such as QBIC [3], MARS [19], Virage [1], Photobook [18], VisualSEEk [24], PicToSeek [4] and so on.

Despite these research efforts, the retrieval performance of current CBIR systems is still relatively unsatisfactory. CBIR systems interpret the user information needs based on a set of low-level visual features (color, texture, shape) extracted from the images. However, these features may not correspond to the user interpretation and understanding of image contents. Thus, a semantic gap exists between the high-level concepts and the low-level features in CBIR. Furthermore, user semantic interpretation depends on individual subjectivity, and may change progressively throughout the searching process. Therefore, relevance feedback (RF), as an interactive mechanism, has been introduced to address these problems [8, 20, 31]. The main idea is that the user is incorporated into the retrieval systems to provide his/her evaluation on the retrieval results. This enables the systems to learn from the feedbacks in order to retrieve a new set of images that better satisfy the user information requirement.

Relevance feedback can be considered as a learning problem. The systems acquire knowledge through learning from the users' feedbacks, and progressively improve the retrieval performance. Many relevance feedback algorithms have been adopted in CBIR systems and demonstrated considerable performance improvement. Query refinement [15, 19] and feature re-weighting [7, 20] are two widely used relevance feedback methods in CBIR. Query refinement tries to reach an optimal query point by moving it towards relevant images and away from the irrelevant ones. This technique has been implemented in many CBIR systems. The best-known implementation is the MARS (Multimedia Analysis and Retrieval System) [19]. Re-weighting aims at emphasizing the feature's dimensions that help to retrieve relevant images, while de-emphasizing those that hinder this process by updating the weights of the feature vectors. It uses heuristic formulation with empirical parameter adjustment. To address this issue, the MindReader retrieval system proposes an optimization procedure to minimize the distances between the query and all the relevant images [9]. The optimal new query turns out to be the weighted average of the relevant images with the optimal distance metric being the Mahalanobis distance. Rui et al. further improves the MindReader approach by developing a unified framework that combines the optimal query estimation and weighting functions [21].

The statistical relevance feedback approaches incorporate the users' feedbacks to update the probability distribution of all images in the database [2, 28]. Bayesian classifier has been proposed that treats positive and negative feedback samples with different strategies [25]. Positive examples are used to estimate a Gaussian distribution that represents the desired images for a given query, while the negative examples are used to modify the ranking of the retrieved candidates. Artificial neural networks have also been adopted in relevance feedback due to their learning and generalization capability [11, 12, 16, 17]. Another popular feedback method in CBIR is centered on support vector machine (SVM). It utilizes the samples near the boundary as the support vectors to minimize the classification error. Tong et al. proposes an SVM active learning algorithm for image retrieval in [27]. To address the small sample problem, Discriminant-Expectation Maximization (D-EM) algorithm has also been introduced to incorporate the unlabeled data during the training process [29]. The results are promising, but the computation complexity can be significant for large databases.

Conventional relevance feedback restricts users to binary labeling of feedback images, namely, the images are determined as either "fully relevant" or "totally irrelevant" [7, 15, 19, 28]. This feedback process is based on hard-decision as to whether the retrieval results satisfy the user requirement. It does not reflect the nature of user interpretation and understanding of images, which tends to be uncertain or imprecise due to perceptual subjectivity. It is, therefore, inappropriate to describe the feedback decision by binary crisp logic without considering the degree of relevance. On the other hand, multi-level labeling categorizes the positive and negative examples into several discrete levels of (ir)relevance [20, 21, 30]. The technique, however, is both inconvenient as well as ambiguous as the users need to classify an image into one of the multiple levels. This conflicts with the uncertainty embedded in human perception. A user is more inclined towards using expressions such as "this image is more or less relevant" or "this image is more relevant than that one".

Radial basis function (RBF) network has been used to determine the nonlinear relationship between features so that more accurate similarity comparison between images can be supported [12]. A single radial basis function-based relevance feedback (SRBF-RF) method has been adopted to model global characterization of image similarity [16]. It utilizes a single model based on Gaussian-shaped RBF. An adaptive radial basis function network (ARBFN) model has also been proposed for interactive image retrieval [17]. It characterizes the query by multi-class models, an inherent strategy of local modeling, and associates those relevant (positive) samples as the models. The irrelevant (negative) samples are used to modify the models such that the models will be moved slightly away from the irrelevant samples. Nevertheless, this method is heuristic as there is no adequate learning process to optimize the underlying network parameters since a single-pass strategy is adopted. Further, it utilizes a binary labeling scheme for relevance feedback, and does not take into account the user interpretation of image similarity.

Taken into account the above-mentioned problems, a fuzzy relevance feedback concept is proposed, which integrates the users' imprecise interpretation of image similarity into the notion of relevance feedback. In addition to the traditional labeling of relevant and irrelevant decisions, a third option called fuzzy decision is incorporated into the feedback mechanism. Users are then allowed to describe some images as fuzzy, if there are ambiguities in the retrieved images. Based on the users' feedbacks, a fuzzy radial basis function network (FRBFN) is proposed to learn the different degrees of relevance embedded in the user interpretation of image similarity. The relevance of the fuzzy feedbacks is evaluated using a properly chosen fuzzy membership function. The network parameters undergo an efficient gradient-descent-based learning procedure by minimizing a cost function. The trained FRBFN is then used in the next session to retrieve the images.

This chapter is organized as follows. Section 2 presents the proposed notion of fuzzy relevance feedback. In Section 3, we describe the FRBFN construction, the formulation of fuzzy membership function, and the learning process. Experimental results using the proposed method are discussed and compared with other techniques in Section 4. Finally, concluding remarks are given in Section 5.

2 Fuzzy Relevance Feedback

2.1 Relevance Judgement in CBIR

The relevance feedback process in image retrieval involves the users' decision-making on the retrieval results. Typically, this is achieved by performing binary labeling where the users have to make a binary decision of "relevant" (positive) and "irrelevant" (negative) for each retrieved image. The system then learns from the users' feedbacks to further improve its performance.

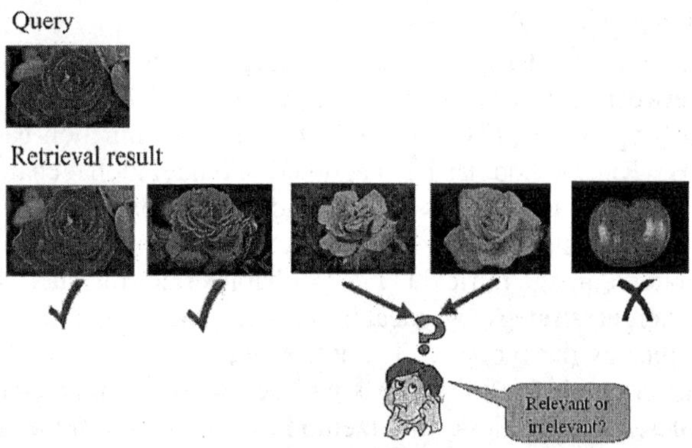

Fig. 1. Scenario of user ambiguity occurring during the feedback process

User interpretation of visual contents is, at times, uncertain or ambiguous as it tends to vary with respect to different users and circumstances. In other words, the notion of image similarity is fuzzy or nonstationary. An example of such a scenario occurring in image retrieval can be seen in Fig. 1, where the user submits the query of a red flower with the intention to locate some flowers and in particular red flowers. The retrieval results contain red flowers, yellow flowers and other objects. Obviously, the user will mark the red flowers as relevant, the non-flower images as irrelevant. However, uncertainties arise in the cases of yellow flowers since they satisfy, up to a certain extent, the information need of the user. If a hard-decision has to be made on the yellow flowers, the user will face the dilemma whether to classify them as either fully relevant or totally irrelevant. It reflects the uncertainty embedded in the decision-making process. Such a decision can affect the outcome of the subsequent retrieval results. If the user marks the yellow flowers as relevant, what he/she really implies is that they satisfy his/her information need up to a certain extent. Therefore, the matching degree cannot be specified by the user with binary labeling.

2.2 Proposed Notion of Fuzzy Relevance Feedback

Uncertainty arises naturally and cannot be handled adequately by binary or multi-level labeling. The users' notion of image similarity is inherently fuzzy. It is difficult to describe it by crisp logic. Fuzzy logic is a superset of conventional logic that has been extended to handle the concept of partial truth, which refers to truth-values that fall between "completely true" and "completely false". Therefore, fuzzy logic can be incorporated into the interactive relevance feedback session to address the ambiguity which is often reflected in the user interpretation of visual contents.

We propose a fuzzy relevance feedback concept which integrates the fuzzy interpretation into the notion of relevance feedback. In addition to the classical labeling of relevant and irrelevant decisions, a third option called fuzzy decision is incorporated into the feedback process. Under this scheme, the retrieved images can fall under one of the three classes: relevant, irrelevant or fuzzy. As opposed to making a hard-decision, it allows a soft-decision to be made with respect to the retrieval results in the form of fuzzy feedbacks. Thus the proposed scheme reconciles the dilemma of binary or multi-level labeling by employing soft-decision instead of hard-decision on the retrieval results. This resembles more closely to the users' decision-making process in interactive image retrieval. It provides a natural and flexible way in expressing the users' preferences and models the interpretation in a gradual and qualitative manner. We utilize a continuous fuzzy membership function to model these fuzzy feedbacks. Different images under the fuzzy label are mapped into suitable relevance scores. With the users' feedbacks, an FRBFN is constructed and an efficient learning algorithm is developed to estimate the underlying parameters based on the scores.

3 Fuzzy Radial Basis Function Network (FRBFN)

3.1 FRBFN Structure

In interactive CBIR, global modeling of image similarity may not adequately address the local data information defined by the current query. To exploit the multiple local properties of image relevance, it is more desirable to adopt local modeling strategy. Thus the local information distributed among multiple classes can be sufficiently exploited to obtain better generalization. Further, the learning rate of the retrieval system should be fast since the users often interact with the system in real time.

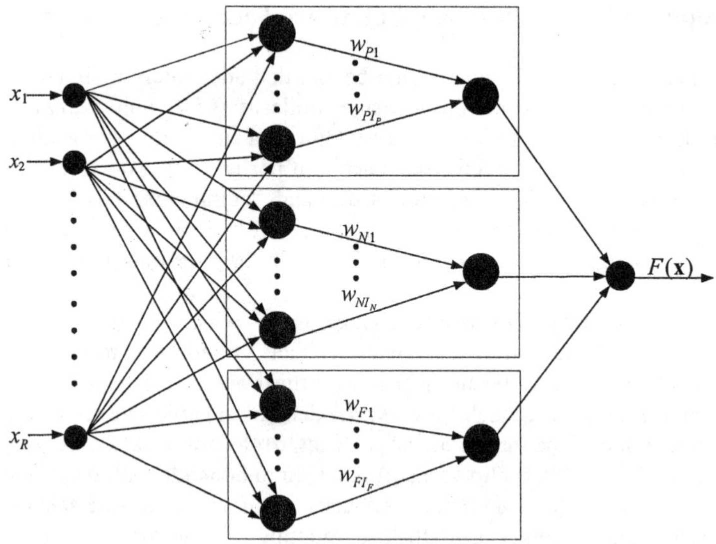

Fig. 2. Architecture of FRBFN

Due to its fast learning speed, simple structure, local modeling and global generalization power, RBF network is adopted in our system to learn the input-output mapping [5]. We propose an FRBFN to progressively learn the user perception of interested images through continual feedbacks. During each feedback iteration, all of the positive, negative and fuzzy samples accumulated over previous feedback sessions are used as the training samples to dynamically create the FRBFN. The architecture of FRBFN is shown in Fig. 2. It consists of an input layer, a hidden layer and an output layer. The input data to FRBFN is R-dimensional feature vectors, $\mathbf{x} = [x_1, x_2, \ldots, x_R]^T$. They are connected to the hidden layer which is constructed from the relevant, irrelevant, and fuzzy samples. The output layer consists of a single unit whose output value $F(\mathbf{x})$ is the linear combination of all the responses from each RBF unit. $\{w_{Pi}\}_{i=1}^{I_P}$, $\{w_{Ni}\}_{i=1}^{I_N}$, and $\{w_{Fi}\}_{i=1}^{I_F}$ are the sets of output connection weights associated with the positive, negative and fuzzy samples, respectively. I_P, I_N and I_F are the number of RBF units for the positive, negative and fuzzy samples, respectively. The detailed network structure and parameters updating algorithm will be described in the following subsections.

3.2 Hierarchical Clustering for FRBFN Center Selection

In some RBF learning methods such as [17], each RBF center is associated with a positive sample. The method models the query by multi-class structure by using one center for each positive sample. The negative samples are used to modify the centers such that the centers will be moved slightly away from the negative samples. Only a single-pass training is used in this method. However, it is heuristic as there is inadequate learning process to optimize the underlying network parameters. Therefore, taking into account the real time requirement and the local clustering nature of the samples, a more systematic and efficient approach to choose the centers is required. We propose to use clustering algorithm to determine the FRBFN centers. The resulting small number of effective centers obtained by clustering can provide less computational complexity in online feedback learning. The accumulated feedback samples are separated into three groups: relevant, irrelevant, and fuzzy. The objective here is to cluster each group of samples individually to initialize the hidden neurons of the FRBFN. Hierarchical clustering, which is among the best known unsupervised methods, is adopted. Single linkage clustering, also known as the nearest neighbor technique is used in our system [10].

Let $\{\mathbf{x}_i\}_{i=1}^{n}$ be the samples to be clustered, single linkage clustering proceeds by fusing the samples into groups. The distance measure between two clusters is defined as the smallest distance between samples in their respective clusters:

$$d_{\min}(D_i, D_j) = \min_{\substack{\mathbf{x} \in D_i \\ \mathbf{x}' \in D_j}} \left\| \mathbf{x} - \mathbf{x}' \right\| \tag{1}$$

where $\mathbf{x}$ and $\mathbf{x}'$ are the samples in cluster D_i and D_j, respectively.

The clustering algorithm can be summarized as follows [10]:

1. Initialization.

   The n samples $\{\mathbf{x}_i\}_{i=1}^{n}$ are placed in n singleton clusters $\{D_i\}_{i=1}^{n}$.

2. Hierarchical cluster tree creation.

   Link pairs of clusters that are the closest together. These newly formed clusters are then linked to other clusters to create bigger clusters until all the samples are linked together into a hierarchical cluster tree.

3. Cluster creation.

   Construct clusters by cutting off the hierarchical tree, where clusters are formed when inconsistent values are greater than a predefined threshold. The inconsistency value for each link of the hierarchical cluster tree characterizes each link by comparing its height with the average height of other links at the same level of the hierarchy. The height of each link represents the distance between the two clusters being connected.

3.3 FRBFN Module

After clustering, three sets of separate clusters are obtained, relevant, irrelevant and fuzzy clusters. Let X be the set of all training samples. The representations for the cluster formation of X are given as:

$$X = P \cup N \cup F \tag{2}$$

$$P = \bigcup_{i=1}^{I_P} P_i \quad N = \bigcup_{j=1}^{I_N} N_j \quad F = \bigcup_{k=1}^{I_F} F_k \tag{3}$$

where P, N, F denote positive, negative, and fuzzy clusters, respectively. P_i, N_j, and F_k are the ith positive cluster, jth negative cluster, and kth fuzzy cluster, respectively. The initial RBF centers are determined as the centroid of each cluster. The Gaussian function is selected as the basis function, defined by:

$$f(\mathbf{x}, \mathbf{v}_{\alpha i}, \sigma_{\alpha i}) = \exp\left(-\frac{(\mathbf{x} - \mathbf{v}_{\alpha i})^{\mathrm{T}} \Lambda (\mathbf{x} - \mathbf{v}_{\alpha i})}{2(\sigma_{\alpha i})^2} \right), \quad i = 1, 2, \ldots, I_\alpha \tag{4}$$

where $\mathbf{x} \in \Re^R$ is the input feature vector. $\alpha \in \{P, N, F\}$ is the cluster type, i and $I_\alpha \in \{I_P, I_N, I_F\}$ are the cluster index and the number of clusters with cluster type α, respectively. $\mathbf{v}_{\alpha i}$ is the RBF center which is the centroid of the ith cluster within cluster type α, $\sigma_{\alpha i}$ is the corresponding RBF width. $\Lambda = diag[\frac{1}{\sigma_1}, \frac{1}{\sigma_2}, \ldots, \frac{1}{\sigma_R}]$ is a diagonal matrix that denotes the relative importance of different feature components with σ_r, $r = 1, 2, \ldots, R$ representing the standard deviation of all relevant samples along the rth axis. This assignment scheme is based on the observation that consistent values of a particular feature component for all the relevant samples suggest its reliability in determining image similarity [20]. Therefore, the feature components with consistent values are assigned greater weights, as the inverse of σ_r.

The FRBFN output $F(\mathbf{x})$ for an input feature vector of a particular image $\mathbf{x}$, is defined as:

$$F(\mathbf{x}) = \sum_{\alpha \in \{P,N,F\}} \sum_{i=1}^{I_\alpha} w_{\alpha i} f(\mathbf{x}, \mathbf{v}_{\alpha i}, \sigma_{\alpha i}) \tag{5}$$

where $w_{\alpha i} \in \{w_{Pi}, w_{Ni}, w_{Fi}\}$ is the network output connection weight.

3.4 Relevance Membership Estimation for Fuzzy Feedbacks

In this study, relevance feedback is implemented as an online supervised learning procedure by adjusting the parameters (center, width and weight) of the network using gradient-descent optimization. To tune the parameters, we need to define the desired network output value $Y(\mathbf{x})$ for each training sample in the first place. The desired network output value is set to 1 and 0 for the training sample $\mathbf{x}$ associated with a positive and negative sample, respectively. In our discussion, we assume the fuzzy images are partially relevant, that is, they are either semantically similar to the users' target image or contain certain features desired by the users. Thus, the desired output value for a fuzzy sample reflects the degree of relevance embedded in the user interpretation of image similarity. Our goal, therefore, is to find a mapping $g(\mathbf{x}): \Re^R \to [0,1]$ that assigns a proper desired output value in the range of $[0,1]$ to each fuzzy feedback sample.

We propose to choose an appropriate fuzzy membership function to estimate the output mapping. Since clustering has been performed on each positive and negative class separately to get multiple clusters per class, the obtained clusters in each class can be employed to generate the membership value of a fuzzy sample. Intuitively, the closer a fuzzy sample to the positive cluster, the higher its degree of relevance. In contrast, the closer a fuzzy sample to the negative cluster center, the lower its degree of relevance. Based on this assumption, the exponential function is selected as the membership function:

$$g(\mathbf{x}) = \exp\left[-a \cdot \frac{\min\limits_i (\mathbf{x} - \mathbf{v}_{Pi})^{\mathrm{T}} \Sigma^{-1} (\mathbf{x} - \mathbf{v}_{Pi})}{\min\limits_j [(\mathbf{x} - \mathbf{v}_{Nj})^{\mathrm{T}} \Sigma^{-1} (\mathbf{x} - \mathbf{v}_{Nj})]} \right] \tag{6}$$

where $\mathbf{v}_{Pi}$ and $\mathbf{v}_{Nj}$ denote the centroid of the ith positive and the jth negative cluster, respectively. $\min\limits_i (\mathbf{x} - \mathbf{v}_{Pi})^{\mathrm{T}} \Sigma^{-1} (\mathbf{x} - \mathbf{v}_{Pi})$ and $\min\limits_j [(\mathbf{x} - \mathbf{v}_{Nj})^{\mathrm{T}} \Sigma^{-1} (\mathbf{x} - \mathbf{v}_{Nj})]$ represent the distance between $\mathbf{x}$ and the closest positive and negative cluster centers, respectively. a is a scaling factor, and Σ is the covariance matrix of all the positive and negative training samples. When the number of training examples is smaller than the dimensionality of the feature space, the singularity issue arises. A statistically viable solution is to add a regularization term on the diagonal of the sample covariance matrix before the matrix inversion. This process is also known as a "linear shrinkage estimator" [30]:

$$\hat{\Sigma} = (1-\mu)\Sigma + \frac{\mu}{R} tr[\Sigma]\mathbf{I} \tag{7}$$

where $tr[\Sigma]$ denotes the trace of Σ. R is the dimension of the feature space, and $0 < \mu < 1$ controls the amount of shrinkage toward the identity matrix $\mathbf{I}$.

Therefore, the desired network output value can be summarized as follows:

$$Y(\mathbf{x}) = \begin{cases} 0 & \mathbf{x} \in N \\ 1 & \mathbf{x} \in P \\ g(\mathbf{x}) & \mathbf{x} \in F \end{cases} \tag{8}$$

3.5 Gradient-Descent Optimization

After the FRBFN is constructed, the network parameters undergo a supervised learning process. A natural candidate for such a process is error-correction learning, which is most conveniently implemented using a gradient-descent procedure [5]. The cost function E at each feedback iteration is defined as:

$$E = \frac{1}{2}\sum_{j=1}^{M} e_j^2 = \frac{1}{2}\sum_{j=1}^{M}\left(Y(\mathbf{x}_j) - F(\mathbf{x}_j)\right)^2 \tag{9}$$

where M is the total number of training samples, and e_j is the error signal for the jth training sample $\mathbf{x}_j$. $F(\mathbf{x}_j)$ and $Y(\mathbf{x}_j)$ represent the actual and desired network output value for $\mathbf{x}_j$, respectively, which have been discussed in previous sections.

By minimizing the cost function over all the training samples through iterative training using gradient-descent optimization, we can update the set of the FRBFN parameters $\theta = \{w_{\alpha i}, \mathbf{v}_{\alpha i}, \sigma_{\alpha i} \mid \alpha \in \{P, N, F\},\ i = 1, 2, \ldots, I_\alpha\}$:

$$\theta = \arg\min_{\theta \in \Theta}(E) = \arg\min_{\theta \in \Theta}\left(\frac{1}{2}\sum_{j=1}^{M}\left(Y(\mathbf{x}_j) - F(\mathbf{x}_j)\right)^2\right) \tag{10}$$

where Θ is the solution space of the parametric vector θ.

The mathematical formulations of parameter updating are summarized as follows:

1. Weight estimation at the kth iteration:

$$\frac{\partial E(k)}{\partial w_{\alpha i}(k)} = -\sum_{j=1}^{M} e_j(k) f(\mathbf{x}_j, \mathbf{v}_{\alpha i}(k), \sigma_{\alpha i}(k)) \tag{11}$$

$$w_{\alpha i}(k+1) = w_{\alpha i}(k) - \eta_1 \frac{\partial E(k)}{\partial w_{\alpha i}(k)} \tag{12}$$

2. Center estimation at the kth iteration:

$$\frac{\partial E(k)}{\partial \mathbf{v}_{\alpha i}(k)} = -w_{\alpha i}(k)\sum_{j=1}^{M} e_j(k) f(\mathbf{x}_j, \mathbf{v}_{\alpha i}(k), \sigma_{\alpha i}(k)) \frac{\Lambda(\mathbf{x}_j - \mathbf{v}_{\alpha i}(k))}{(\sigma_{\alpha i}(k))^2} \tag{13}$$

$$\mathbf{v}_{\alpha i}(k+1) = \mathbf{v}_{\alpha i}(k) - \eta_2 \frac{\partial E(k)}{\partial \mathbf{v}_{\alpha i}(k)} \tag{14}$$

3. Width estimation at the kth iteration:

$$\frac{\partial E(k)}{\partial \sigma_{\alpha i}(k)} = -w_{\alpha i}(k)\sum_{j=1}^{M} e_j(k) f(\mathbf{x}_j, \mathbf{v}_{\alpha i}(k), \sigma_{\alpha i}(k)) \frac{(\mathbf{x}_j - \mathbf{v}_{\alpha i}(k))^{\mathrm{T}} \Lambda(\mathbf{x}_j - \mathbf{v}_{\alpha i}(k))}{(\sigma_{\alpha i}(k))^3} \quad (15)$$

$$\sigma_{\alpha i}(k+1) = \sigma_{\alpha i}(k) - \eta_3 \frac{\partial E(k)}{\partial \sigma_{\alpha i}(k)} \quad (16)$$

4. Repeat steps 1–3 until convergence or a maximum number of training iterations is reached.

The term $e_j(k)$ is the error signal for the jth training sample $\mathbf{x}_j$ at the kth iteration. η_1, η_2 and η_3 are different learning parameters for $w_{\alpha i}$, $\mathbf{v}_{\alpha i}$, and $\sigma_{\alpha i}$, respectively.

Unlike the back-propagation algorithm, the gradient-descent approach described here does not involve error back-propagation. Thus, it requires less training time to converge when compared with other neural networks such as multilayer perception network (MLPN). This ensures fast online learning to meet the real time requirement.

4 Experimental Results

4.1 Image Database and Feature Extraction

The image database used in this experiment contains 10,000 natural images of 100 different categories obtained from the Corel Gallery product. These categories are predefined by Corel Photo Gallery based on their semantic concepts. It covers a wide variety of subjects, as shown in Fig. 3. Color histogram [26], color moments [14] and color auto-correlogram [7] are used to represent the color feature, while Gabor wavelet [13] and wavelet moments [23] are used to represent the texture feature. The details of these feature descriptors are given in Table 1.

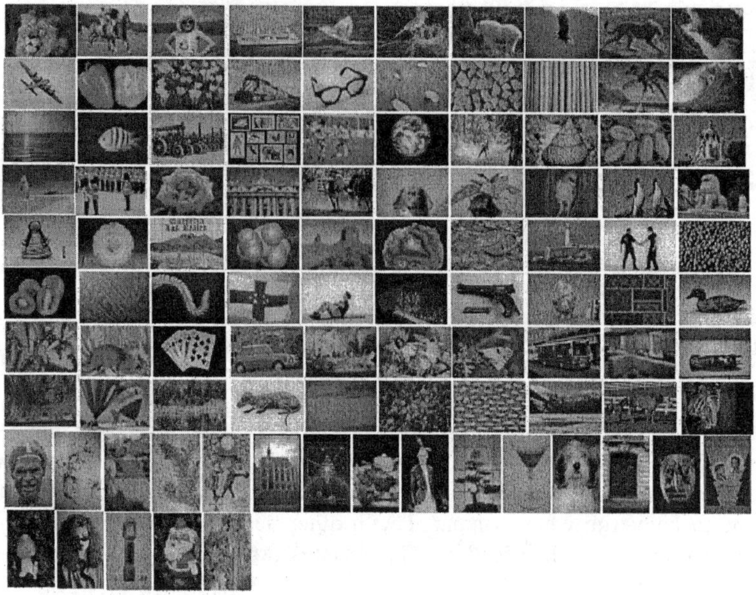

Fig. 3. Selected sample images in the database

Table 1. Extracted Features Used in the Experiment

Features	Description	Dimension
Color histogram	HSV space is chosen, each H, S, V component is uniformly quantized into 8, 2 and 2 bins respectively	32
Color auto-correlogram	Choose D_8 distance (chessboard distance) as the distance measure: $D_8(p,q) = \max(\lvert p_x - q_x \rvert, \lvert p_y - q_y \rvert)$ which is the maximum of the distance in the x direction and the y direction. Quantize the image into 4*4*4=64 colors in the RGB space	64
Color moments	The first two moments (mean and standard deviation) from the three R, G , B color channels are extracted	6
Gabor wavelet	Gabor wavelet filters spanning four scales: 0.05, 0.1, 0.2 and 0.4 and six orientations: $\theta_0 = 0$, $\theta_{n+1} = \theta_n + 6/\pi$ are applied to the image. The mean and standard deviation of the Gabor wavelet coefficients are used to form the feature vector	48
Wavelet moments	Applying the wavelet transform to the image with a 3-level decomposition, the mean and the standard deviation of the transform coefficients are used to form the feature vector	20

4.2 Performance Evaluation

In our experiment, we use both objective and subjective measures to evaluate the performance of the proposed FRBFN method, and compare it with the ARBFN method in [17]. When there is no ambiguity in the user perception, or in other words no fuzzy feedbacks, objective performance measure based on the predefined ground truth is used. That is, the retrieved images are judged to be relevant if they come from the same category as the query. Precision versus recall graph, a standard performance evaluation measure in information retrieval, is adopted in our experiment. Precision and recall are defined as follows [22]:

$$\text{Precision}=\frac{\text{number of relevant images retrieved}}{\text{total number of images retrieved}} \tag{17}$$

$$\text{Recall}=\frac{\text{number of relevant images retrieved}}{\text{total number of relevant images in the database}} \tag{18}$$

100 queries with one from each category are selected for evaluation. For each query, the top 25 retrieved images are used for relevance feedback. Precision is averaged over all queries. The average precision versus recall graph after 5 feedback iterations is shown in Fig. 4. From the precision-vs-recall graph, we observe that the FRBFN method provides higher recall rate than the ARBFN method for the same precision level. It also offers higher precision rate for the same recall level. Thus the FRBFN method achieves better retrieval performance than the ARBFN method, which indicates the superiority of the proposed learning approach.

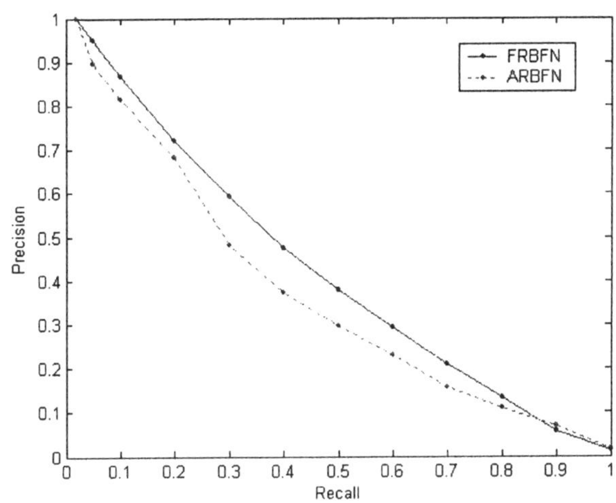

Fig. 4. Average precision versus recall graph (after 5 feedback iterations)

In addition, we use another measure called retrieval accuracy (RA) to evaluate the retrieval performance [6, 25]:

$$\text{Retrieval accuracy} = \frac{\text{relevant images retrieved in top } T \text{ returns}}{T} \quad (19)$$

where T is the number of retrieved images shown to the user for feedback.

The comparison of the retrieval performance using the FRBFN method and the ARBFN method is given in Fig. 5. The retrieval accuracy is averaged over 100 test queries. Seven iterations of feedbacks are recorded. Based on the results, we observe that the FRBFN method consistently achieves higher retrieval accuracy than the ARBFN method in both the top 25 and 50 results. The retrieval accuracy of the FRBFN method increases quickly in the initial stage. After the first feedback iteration, FRBFN method achieves an improvement of 25.6% and 25.7% from the initial value in the top 25 and 50 results, respectively. This is a desirable feature, since the user can obtain significant improvement on the retrieval results quickly. After seven iterations, the retrieval accuracy obtained is 80.0% (FRBFN), 75.0% (ARBFN) in the top 25 results, and 70.5% (FRBFN), 67.0% (ARBFN) in the top 50 results. Further, it is observed that to achieve a specific retrieval accuracy, the FRBFN method requires less number of iterations when compared to the ARBFN method.

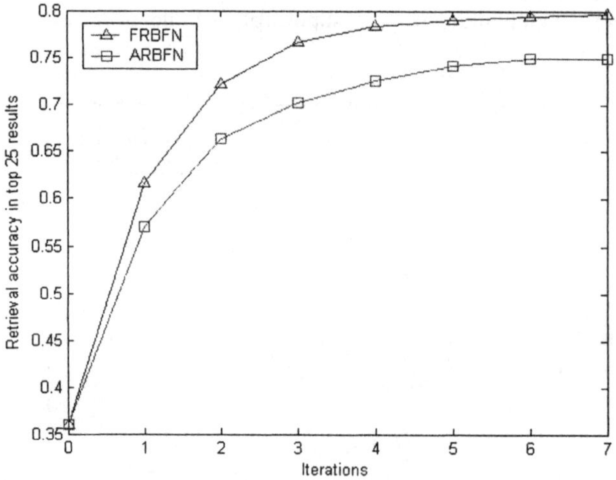

(a)

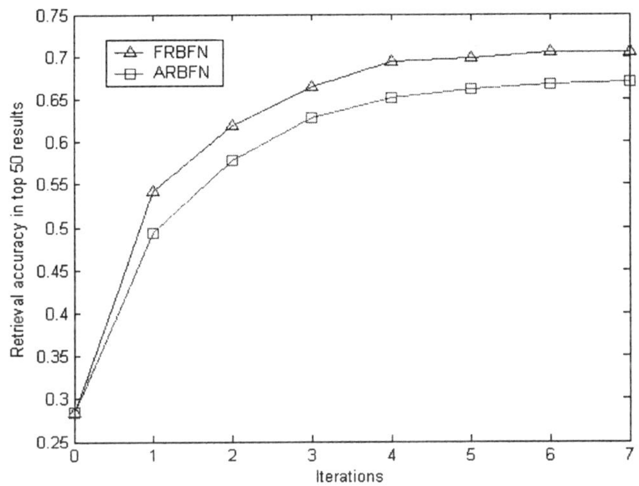

(b)

Fig. 5. Performance comparison of FRBFN and ARBFN based on ground truth. (a) Retrieval accuracy in top 25 results; (b) Retrieval accuracy in top 50 results

When there are ambiguities in the feedback images, subjective test is used to evaluate the system performance. Six users are invited to evaluate the retrieval system. A total of 180 queries are used for evaluation. We define the following performance measure, total retrieval accuracy (TRA), and relevant retrieval accuracy (RRA):

$$\text{Total retrieval accuracy} = \frac{\text{relevant and fuzzy images retrieved in top } T \text{ returns}}{T} \tag{20}$$

$$\text{Relevant retrieval accuracy} = \frac{\text{relevant images retrieved in top } T \text{ returns}}{T} \tag{21}$$

Since fuzzy samples satisfy the user information need up to a certain extent, they are also part of the users' desired images. Therefore, TRA is introduced to incorporate the users' fuzzy feedbacks. TRA and RRA can be considered as the upper and lower bounds of effective retrieval accuracy. Together, they give an overall idea on the performance of the FRBFN method.

The comparison of the retrieval performance using the FRBFN and ARBFN methods is given in Fig. 6. The zeroth iteration denotes the users' relevance judgement on the initial retrieved images based on k-nearest neighbors (K-NN) search. The retrieval performance is averaged over all queries and users. We observe that our FRBFN method provides better retrieval performance than the ARBFN method. The retrieval accuracy increases with respect to the number of feedback iterations. After seven iterations, the retrieval accuracy obtained is 87.2% (TRA), 85.3% (RRA), 77.3% (ARBFN) in the top 25 results, and 73.1%

(TRA), 70.6% (RRA), 67.4% (ARBFN) in the top 50 results. The results are consistent with the objective performance evaluation that the FRBFN method outperforms the ARBFN method.

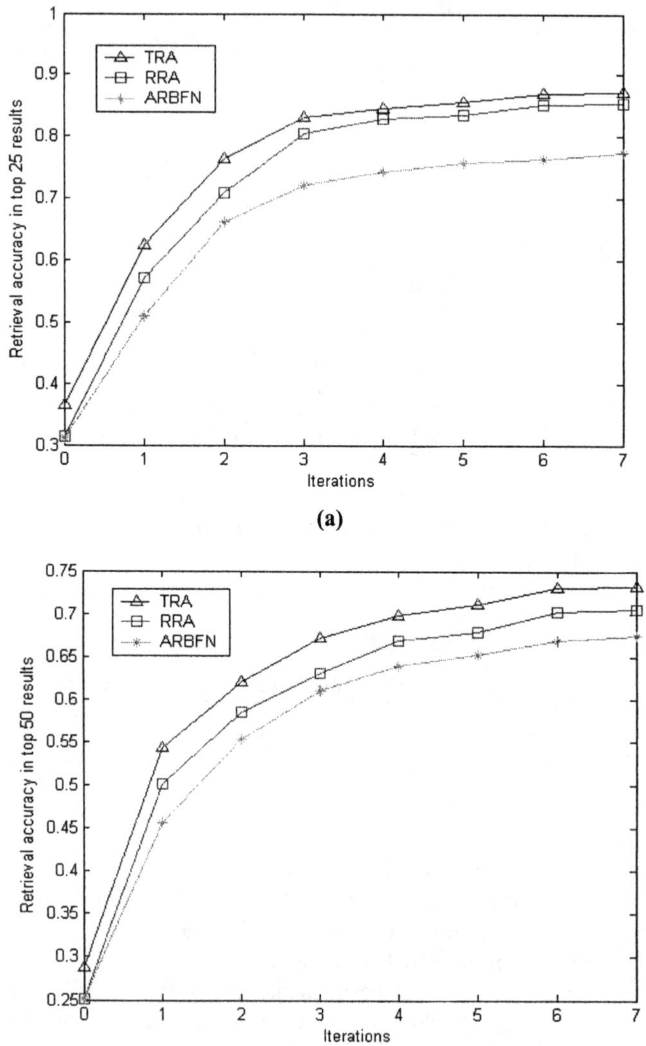

(a)

(b)

Fig. 6. Performance comparison of FRBFN and ARBFN based on user subjectivity. (a) Retrieval accuracy in top 25 results; (b) Retrieval accuracy in top 50 results

Considering the real time requirement for interactive CBIR, we compare the retrieval time for both the FRBFN and ARBFN methods. In our experiments running on a Pentium 4 2.4GHz PC with MATLAB programming, the retrieval time using the FRBFN method is comparable with the ARBFN method, ranging from 5 to 10 seconds for the top 25 results. The ARBFN method adopts a single-pass learning strategy to achieve learning. Nevertheless, the procedure is inadequate in achieving good generalization. In contrast, the FRBFN method develops robust learning scheme to estimate the underlying network parameters.

Fig. 7 gives an example of a retrieval session to illustrate the effectiveness of the FRBFN method.

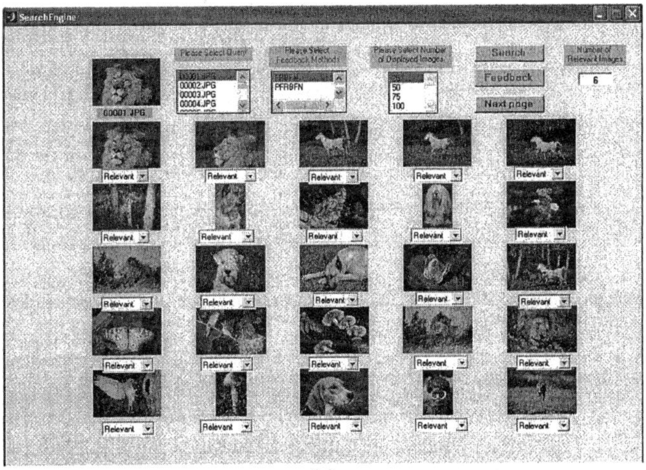

(a)

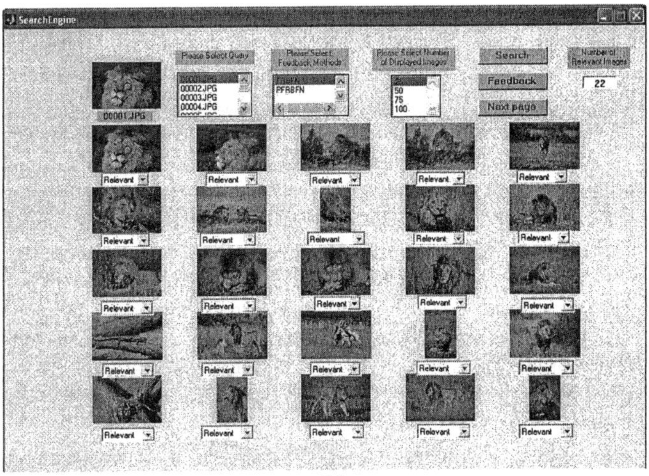

(b)

Fig. 7. Retrieval results (query image is on the left corner). (a) Initial retrieval results without feedback; (b) Retrieval results using FRBFN method after 1 feedback iteration

5 Conclusion

This chapter presents the framework of fuzzy relevance feedback and the corresponding FRBFN to integrate the users' imprecise interpretation of image similarity into interactive CBIR systems. During each feedback iteration, the FRBFN is constructed dynamically using hierarchical clustering and trained by gradient-descent optimization to improve the retrieval performance. In contrast to conventional relevance feedback approaches that are based on binary or multi-level labeling, our method provides a natural and flexible way to express the users' preferences. It reconciles the dilemma faced by traditional relevance feedback schemes where a hard-decision on the relevance of the retrieved images has to be made. Experimental results show that the new method is effective in addressing different users' information needs in interactive image retrieval systems.

References

1. Amarnath G, Ramesh J (1997) Visual information retrieval. Communications of ACM, vol. 40, no.5, pp.70–79
2. Cox IJ, Miller ML, Minka TP, Papathomas TV, Yianilos PN (2000) The Bayesian image retrieval system, PicHunter: Theory, implementation, and psychophysical experiments. IEEE Trans. on Image Processing, vol. 9, no. 1, pp. 20–37
3. Flickher M, Sawhney H, Niblack W, Ashley J, Huang Q, Dom B, Gorkani M, Hafner J, Lee D, Petkovic D, Steele D, Yanker P (1995) Query by image and video content: The QBIC system. IEEE Computer, vol. 28, no. 9, pp. 23–32
4. Gevers T, Smeulders AWM (2000) PicToSeek: Combining color and shape invariant features for image retrieval. IEEE Trans. on Image Processing, vol. 9, pp. 102–119
5. Haykin S (1999) Neural networks: A comprehensive foundation. Upper Saddle River, NJ: Prentice-Hall
6. He XF, King O, Ma WY, Li MJ, Zhang HJ (2003) Learning a semantic space from user's relevance feedback for image retrieval. IEEE Trans. on Circuits and Systems for Video Technology, vol. 13, pp. 39–48
7. Huang J, Kumar SR, Metra M (1997) Combining supervised learning with color correlograms for content-based image retrieval. Proc. of ACM Multimedia, pp. 325–334
8. Huang TS, Zhou XS, Munehiro N, Wu Y, Cohen I (2002) Learning in content-based image retrieval. 2nd Int. Conf. on Development and Learning
9. Ishikawa Y, Subramanya R (1998) MindReader: Query database through multiple examples. Proc. of Int. Conf. on Very Large Data Bases, New York, USA
10. Jain AK, Murty MN, Flynn PJ (1999) Data clustering: A review. ACM Computing Surveys, Vol. 31, No. 3, pp. 264–323
11. Laaksonen J, Koskela M, Oja E (2002) PicSom–self-organizing image retrieval with MPEG-7 content descriptions. IEEE Trans. on Neural Network, vol. 13, no. 4, pp. 841–853
12. Lee HK, Yoo SI (2001) A neural network-based image retrieval using nonlinear combination of heterogeneous fcatures. International Journal of Computational Intelligence and Applications 1(2): 137–149

13. Manjunath BS, Ma WY (1996) Texture features for browsing and retrieval of image data. IEEE Trans. on Pattern Analysis and Machine Intelligence, vol. 18, pp. 837–842
14. Markus S, Markus O (1995) Similarity of color images. Proc. SPIE Storage and Retrieval for Image and Video Databases
15. Müller H, Müller W, Marchand-Maillet S, McG Squire D (2000) Strategies for positive and negative relevance feedback in image retrieval. Proc. Int. Conf. on Pattern Recognition, Barcelona, Spain
16. Muneesawnag P, Guan L (2001) Interactive CBIR using RBF-based relevance feedback for WT/VQ coded images. Proc. of IEEE Int. Conf. on Acoustics, Speech and Signal Processing, Utah, USA
17. Muneesawnag P, Guan L (2002) Automatic machine interactions for content-based image retrieval using a self-organizing tree map architecture. IEEE Trans. on Neural Networks, vol. 13, no. 4, pp. 821–834
18. Pentland A, Picard R, Sclaroff S (1994) Photobook: tools for content-based manipulation of image databases. Proc. SPIE, vol.2185, pp.34–47
19. Rui Y, Huang TS, Mehrotra S (1997) Content-based image retrieval with relevance feedback in MARS. IEEE Int. Conf. on Image Processing, Washington D.C., USA, pp. 815–818
20. Rui Y, Huang TS, Ortega M, Mehrotra S (1998) Relevance feedback: A power tool for interactive content-based image retrieval. IEEE Trans. on Circuits and Video Technology, vol. 8, no. 5, pp. 644–655
21. Rui Y, Huang TS (2000) Optimizing learning in image retrieval. Proc. of IEEE Int. Conf. on Computer Vision and Pattern Recognition, vol.1, pp. 236–243
22. Salton G, McGill MJ (1982) Introduction to Modern Information Retrieval. New York: McGraw-Hill Book Company
23. Smith JR, Chang SF (1996) Automated binary texture feature sets for image retrieval. Proc. Int. Conf. on Acoustics, Speech and Signal Processing, Atlanta, GA
24. Smith JR, Chang SF (1996) VisualSEEk: a fully automated content based image query system. Proc. ACM Multimedia
25. Su Z, Zhang HJ, Li S, Ma SP (2003) Relevance feedback in content-based image retrieval: Bayesian framework, feature subspaces, and progressive learning. IEEE Trans. on. Image Processing, vol. 12, pp. 924–937
26. Swain M, Ballard D (1991) Color indexing. International Journal of Computer Vision, Vol. 7, No. 1, pp. 11–32
27. Tong S, Chang E (2001) Support vector machine active leaning for image retrieval. Proc. of 9th ACM Conf. on Multimedia, Ottawa Canada
28. Vasconcelos N, Lippman A (1999) Learning from user feedback in image retrieval systems. Proc. of Neural Information Processing Systems, Denver, Colorado
29. Wu Y, Tian Q, Huang TS (2000) Discriminant-EM algorithm with application to image retrieval. Proc. of IEEE Int. Conf. on Computer Vision and Pattern Recognition, South Carolina
30. Zhou XS, Huang TS (2001) Small sample learning during multimedia retrieval using BiasMap. Proc. of Int. Conf. on Computer Vision and Pattern Recognition. Hawaii
31. Zhou XS, Huang TS (2003) Relevance feedback in image retrieval: A comprehensive review. Multimedia System 8(6), pp. 536–544

A Scalable Bootstrapping Framework for Auto-Annotation of Large Image Collections

Tat-Seng Chua and Huamin Feng

School of Computing, National University of Singapore, Singapore 117543

Abstract. Image annotation aims to assign semantic concepts to images based on their visual contents. It has received much attention recently as huge dynamic collections of images/videos become available on the Web. Most recent approaches employ supervised learning techniques, which have the limitation that a large set of labeled training samples is required for effective learning. This is both tedious and time consuming to obtain. This chapter explores the use of a bootstrapping framework to tackle this problem by employing three complementary strategies. First, we train two "view independent" classifiers based on probabilistic SVM using two orthogonal sets of content features and incorporate the classifiers in the co-training framework to annotate regions. Second, at the image level, we employ two different segmentation methods to segment the image into different sets of possibly overlapping regions and devise a contextual model to disambiguate the concepts learned from different regions. Third, we incorporate active learning in order to ensure that the framework is scalable to large image collections. Our experiments on a mid-sized image collection demonstrate that our bootstrapping cum active learning framework is effective. As compared to the traditional supervised learning approach, it is able to improve the accuracy of annotation by over 4% in F_1 measure without active learning, and by over 18% when active learning is incorporated. Most importantly, the bootstrapping framework has the added benefit that it requires only a small set of training samples to kick start the learning process, making it suitable to practical applications.

Keywords: bootstrapping, co-training, image annotation, active learning

1 Introduction

For many years, we have been using automated content-based techniques to model and retrieve large image[1] collections using low-level content features such as color, texture and shape [3, 4, 19, 20]. However, such techniques are of insufficient accuracy for many practical applications. An alternative is to invest large amount of pre-processing efforts in automatically annotating images using keywords or concepts[2] so that users may issue concept-based queries to search for images. Here we define annotation as the process of associating one or more pre-defined

---

[1] All references to images also imply videos.

[2] Throughout this chapter, we liberally use the term concept and keyword interchangeably.

concepts with new images based on their visual contents. Although many useful image collections come with keyword annotations, the annotations are normally incomplete, and there are many more image collections that do not have such annotations. Thus there is a need to develop (semi) automated techniques to annotate images with semantic concepts accurately and completely.

We view image annotation as a classification process aiming to determine if the content of the whole or part of an image can be classified into one of N pre-defined categories or concepts. We therefore call the automated systems that perform the annotation as **classifiers**. In general, the task of annotating images can be divided into three sub-stages of: (a) segmenting images into meaningful units; (b) extracting appropriate features for each unit; and (c) associating these features with text. Here different techniques may be used to divide the images into sub-units in Stage (a), which could simply be the whole image, a fixed size block, or a region. Stage (b) is the feature extraction problem at the sub-unit level, while Stage (c) associates content features of sub-unit with concepts, typically using a learning-based method.

One popular approach to achieve automated annotation is to employ a supervised learning approach to perform Stage (c), such as the use of 2D HMM [23] or SVM [7] methods. The main practical problem with such approaches is that a large set of labeled training samples is needed, and it is very tedious and time-consuming to provide such training samples. Moreover, such learning approach is "passive" and is unable to learn incrementally and adapt to changing domain.

To overcome the problems of supervised learning approaches, we propose a bootstrapping cum active learning approach to perform auto image annotation using three strategies. The first strategy is to perform bootstrapping [1] at the region level to derive the concepts. Bootstrapping aims to use a small set of labeled samples to kick-start the learning process from a large unlabeled corpus. To perform bootstrapping, we need a way for the system to evaluate the quality of new annotated samples. This can be achieved by using the co-training technique [4] in which two "view independent" methods independently confirm the quality of newly annotated samples, and learn from each other's results. This can be accomplished by devising two independent sets of features in Stage (b), and using that to train two independent classifiers in Stage (c). As the aim of this chapter is not to investigate better techniques to model image contents, we consider only the use of two common orthogonal sets of content features – one based on color histogram, and the other on texture and shape features. The use of more advanced features to model image contents is currently being investigated [21].

The second strategy is to employ a contextual model at the image level to disambiguate the set of concepts learned at the image level using the overlapping set of regions generated from two separate image segmentation methods. We employ two separate image segmentation methods as the current segmentation methods are unreliable and unstable. The use of multiple methods helps to

Each pixel takes on a discrete value in the label set $\mathcal{L} = 1, \ldots, M$. The spatial relationship of the sites is determined by a neighborhood system $\mathcal{N} = \{\mathcal{N}_i \mid i \in \mathcal{S}\}$ where $\mathcal{N}_i$ is the set of sites neighboring i. A single site or a set of neighboring sites form a clique denoted by c. $\mathcal{C}$ is the set of all cliques.

There are many different ways the pixels can influence each other through contextual interactions. Such contextual interactions can be modeled as MRFs. According to Markov-Gibbs equivalence [13], an MRF is equivalent to a Gibbs distribution

$$P(x) = Z^{-1} \times e^{-\frac{1}{T}\sum_{c\in\mathcal{C}} V_c(x)} \tag{2}$$

where $V_c(x)$ is the clique potential function, T is a temperature constant, and Z is a normalizing constant.

In the MAP-MRF labeling, the posterior distribution can be computed by using [14]

$$P(x|y) = \frac{p(y|x)P(x)}{p(y)} \tag{3}$$

$p(y)$ is a constant when y is given. $P(x)$ is the prior probability and $p(y|x)$ is the conditional distribution. The prior distribution of the original image which imposes the contextual constraints can be expressed in terms of the MRF clique potentials

$$P(x) = Z^{-1} \times e^{-\frac{1}{T}\sum_{c\in\mathcal{C}} U(x)} \tag{4}$$

where

$$U(x) = \sum_{\{i\in\mathcal{C}\}} V_1(x) + \sum_{\{i,i'\}\in\mathcal{C}} V_2(x_i, x_{i'}) + \cdots \tag{5}$$

In this chapter only pair-site cliques are considered.

In (1) the noise is independent Gaussian noise. The conditional distribution can be expressed in terms of y, x and σ.

$$p(y|x) = \frac{1}{\prod_i^N \sqrt{2\pi\sigma^2}} e^{-U(y|x)} \tag{6}$$

where

$$U(y|x) = \sum_{i\in\mathcal{S}} \frac{(x_i - y_i)^2}{2\sigma^2} \tag{7}$$

Knowing the prior distribution and the conditional distribution, the energy in the posterior distribution is given by [4]

$$E(x) = \sum_{i\in\mathcal{S}} \frac{(x_i - y_i)^2}{2\sigma^2} + \sum_{\{i,i'\}\in\mathcal{C}} V_2(x_i, x_{i'}) \tag{8}$$

in which $\sum_{\{i,i'\}\in\mathcal{C}} V_2(x_i, x_{i'})$ is the pair-site clique potential of the MRF model. Maximizing *a posteriori* is equivalent to find an $\hat{x}$ such that $E(\hat{x})$ is the global minimum.

In order to minimize $E(x)$ in (8), proper MRF model has to be chosen so that appropriate contextual constraints can be posed. Among various MRFs, the multi-level logistic (MLL) model is a simple mechanism for encoding a large class of spatial patterns [14]. In MLL, the pair-site clique potentials take the form

$$V_2(x_i, x_{i'}) = \begin{cases} -\beta_c & \text{if } x_i = x_{i'} \\ \beta_c & \text{otherwise} \end{cases} \tag{9}$$

where β_c is the parameter for cliques of type $c = \{i, i'\}$. However we found that the pair-site potentials are strong constraints. Instead of using (9) we use

$$V_2(x_i, x_{i'}) = -\beta_c * g(x_i, x_{i'}) \tag{10}$$

$$g(x_i, x_{i'}) = 2e^{-|x_i - x_{i'}|/\tau} - 1 \tag{11}$$

where τ is a positive constant.

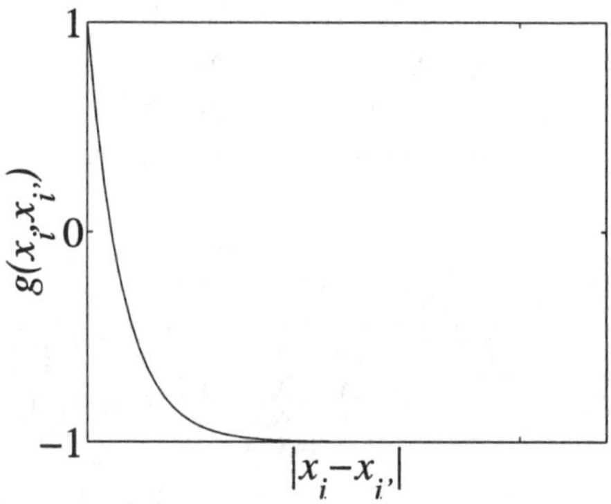

Fig. 1. Graphical representation of $g(x_i, x_{i'})$

$g(x_i, x_{i'})$ in the modified potential function is an exponential function in (-1,1]. Compared to the potential function in the MLL model, the modified potential function allows the pixels to be slightly different from the neighboring pixels. This is logical as most real images have smooth non-uniform regions.

Since image pixels can only take discrete values, the minimization in (8) is combinatorial. The combinatorial optimization can be converted into a constrained optimization in a real space using the continuous relaxation labeling. Let $p_i(I) \in [0,1]$ represent the strength with which label I is assigned to i, the energy with the p variables is given by

$$E(p) = \sum_{i \in \mathcal{S}} \sum_{I \in \mathcal{L}} p_i(I) + \sum_{i \in \mathcal{S}} \sum_{I \in \mathcal{L}} \sum_{i' \in \mathcal{S}, i' \neq i} \sum_{I' \in \mathcal{L}} r_{i,i'}(I, I') p_i(I) p_{i'}(I') \tag{12}$$

where $I = x_i$ and $I' = x_{i'}$, $r_i(I) = V_1(I|y) = (I - y_i)^2/2\sigma^2$ is the single-site clique potential function and $r_{i,i'}(I, I') = V_2(I, I'\ |y) = V_2(I, I')$ is the pair-site clique potential function in the posterior distribution $P(x|y)$.

With such a representation, the combinatorial minimization is reformulated as the following constrained minimization

$$\min_{p} \ E(p) \tag{13}$$

$$\text{subject to } \ C_i(p) = 0 \qquad i \in \mathcal{S} \tag{14}$$

$$p_i(I) \geq 0 \qquad \forall i \in \mathcal{S},\ \forall I \in \mathcal{L} \tag{15}$$

where $C_i(p) = \sum_I p_i(I) - 1 = 0$.

The augmented Lagrange Hopfield (ALH) method can be used to solve the above constrained minimization problem [4]. It uses the augmented Lagrange technique to satisfy the equality constraints of (14) and the Hopfield encoding to impose the inequality constraints of (15). The augmented Lagrange function takes the form

$$L_\beta(p, \gamma) = E(p) + \sum_{k \in \mathcal{S}} \gamma_k C_k(p) + \frac{\beta}{2} \sum_{k \in \mathcal{S}} [C_k(p)]^2 \tag{16}$$

where γ_k are the Lagrange multipliers and $\beta > 0$ is the weight for the penalty term. The final solution p^* is subject to additional constraints: $p_i^*(I) \in \{0, 1\}$.

The ALH method uses the HNN to optimize the energy function. However the HNN is prone to trappings at local minima. In view of this, we propose a new network NCNN to perform the optimization.

3 Noisy Chaotic Neural Network

Let $u_i(I)$ denote the internal state of the neuron (i, I) and $p_i(I)$ denote the output of the neuron (i, I). $p_i(I) \in [0, 1]$ represents the strength that the pixel at location i takes the value I. The NCNN is formulated as follows [12]:

$$p_i^{(t)}(I) = \frac{1}{1 + e^{-u_i^{(t)}(I)/\epsilon}} \tag{17}$$

$$u_i^{(t+1)}(I) = k u_i^{(t)}(I) - z^{(t)}(p_i^{(t)}(I) - I_o) + n^{(t)} + \alpha\Big(\sum_{i'=1, i' \neq i}^{N} \sum_{I'=1}^{M} T_{iI;i'I'} p_{i'}^{(t)}(I') + \mathcal{I}_i(I)\Big) \tag{18}$$

$$z^{(t+1)} = (1 - \beta_z) z^{(t)} \tag{19}$$

$$A[n^{(t+1)}] = (1 - \beta_n) A[n^{(t)}] \tag{20}$$

where
$T_{iI;i'I'}$: connection weight from neuron (i', I') to neuron (i, I);
$\mathcal{I}_i(I)$: input bias of neuron (i, I);
k : damping factor of nerve membrane $(0 \leq k \leq 1)$
α : positive scaling parameter for inputs;
ϵ : steepness parameter of the output function $(\epsilon \geq 0)$;
z : self-feedback connection weight or refractory strength $(z \geq 0)$;
I_o : positive parameter;
n : random noise injected into the neurons;
β_z : positive parameter $(0 < \beta_z < 1)$;
β_n : positive parameter $(0 < \beta_n < 1)$;
$A[n]$: the noise amplitude.

When $n^{(t)}$ in (18) equals to zero, the NCNN becomes TCNN. When $z^{(t)}$ equals to zero, the TCNN becomes similar to the HNN with stable fixed point dynamics. The basic difference between the HNN and the TCNN is that a nonlinear term $z^{(t)}(p_i^{(t)}(I) - I_o)$ is added to the HNN. Since the "temperature" $z^{(t)}$ tends toward zero with time evolution, the updating equations of the TCNN eventually reduce to those of the HNN. In (18) the variable $z^{(t)}$ can be interpreted as the strength of negative self-feedback connection of each neuron, the damping of $z^{(t)}$ produces successive bifurcations so that the neurodynamics eventually converge from strange attractors to a stable equilibrium point [8].

CSA is deterministic and is not guaranteed to settle into a global minimum. In view of this, Wang and Tian [12] added a noise term $n^{(t)}$ in (18). The noise term continues to search for the optimal solution after the chaos of the TCNN disappears.

From (13)-(15) and (18), we obtain the dynamics of the NCNN:

$$u_i^{(t+1)}(I) = ku_i^{(t)}(I) - z^{(t)}(p_i^{(t)}(I) - I_o) + n^{(t)} + \alpha q_i^{(t)}(I) \tag{21}$$

where

$$q_i(I) = -\frac{\partial L_\beta(p, \gamma)}{\partial p_i(I)}$$

Note that the Lagrange multipliers are updated with neural outputs according to $\gamma_k^{(t+1)} = \gamma_k^{(t)} + \beta C_i(p^{(t)})$.

4 Experimental Results

Both artificial image and real image have been used to demonstrate the performance of the NCNN on image denoising.

The artificial image is a circle image of size 256×256 with $M = 4$ gray levels. Its label set is $\mathcal{L} = \{0, 1, 2, 3\}$. Three noisy circle images were generated by adding zero-mean i.i.d. Gaussian noise with standard deviation $\sigma = 0.5$, $\sigma = 0.75$ and $\sigma = 1$ respectively.

The noisy images were set to be the input of the neural networks. After the neural networks were initialized, each neuron was updated using (21) and (17). After all neurons in the neural networks were updated once, γ_k, z and n are updated. The updating scheme is cyclic and asynchronous. When all the neurons are updated once, we call it one iteration. Once the state of a neuron is updated, the new state information is immediately available to other neurons in the network (asynchronous).

The parameters that we used for the NCNN are: $k = 1$, $\epsilon = 0.01$, $\alpha = 0.005$, $I_0 = 0.65$, $z^{(0)} = 0.05$, $n^{(0)} = 0.01$. β is increased from 1 to 50 according to $\beta \leftarrow 1.01\beta$. The decreasing rate of z and n, β_z and β_n are 0.005. For the TCNN and the HNN we use the same parameters as the NCNN except that $n = 0$ for the TCNN, $n = 0$ and $z = 0$ for the HNN. The MRF pair-site clique potential parameter β_c=1.

Table 1 shows the required iteration numbers and the peak signal-to-noise ratio (PSNR) of the denoised images. The higher the PSNR, the better the image quality. The denoised images are shown in Fig. 2-4.

Table 1. Numerical denoising results of the circle image ($PSNR_r$: PSNR of the denoised image, $PSNR_d$: PSNR of the noisy image)

Noise Level		Iterations	$PSNR_r$	$PSNR_d$
$\sigma = 0.5$	HNN	23	35.8127	18.7018
	TCNN	75	38.935	18.7018
	NCNN	199	38.9675	18.7018
$\sigma = 0.75$	HNN	121	30.6347	15.8694
	TCNN	125	33.0543	15.8694
	NCNN	206	33.7814	15.8694
$\sigma = 1$	HNN	226	27.3437	14.7626
	TCNN	137	28.4625	14.7626
	NCNN	151	29.4505	14.7626

The next experiment was conducted on the real image. The real image is the Lena image of size 128 × 128 with $M = 256$ gray levels. Its label set is $\mathcal{L} = \{0, 1, 2, \ldots, 255\}$. Three noisy Lena images were generated by adding zero-mean i.i.d. Gaussian noise with standard deviation of $\sigma = 8$, $\sigma = 16$ and $\sigma = 24$ respectively.

The denoising process of the noisy Lena images is the same as the circle image denoising process. The parameters that we used for the NCNN of the Lena image denoising process are: $k = 1$, $\epsilon = 0.01$, $\alpha = 0.0001$, $I_0 = 0.65$, $z^{(0)} = 0.05$, $n^{(0)} = 0.01$. β is increased from 1 to 50 according to $\beta \leftarrow 1.01\beta$.

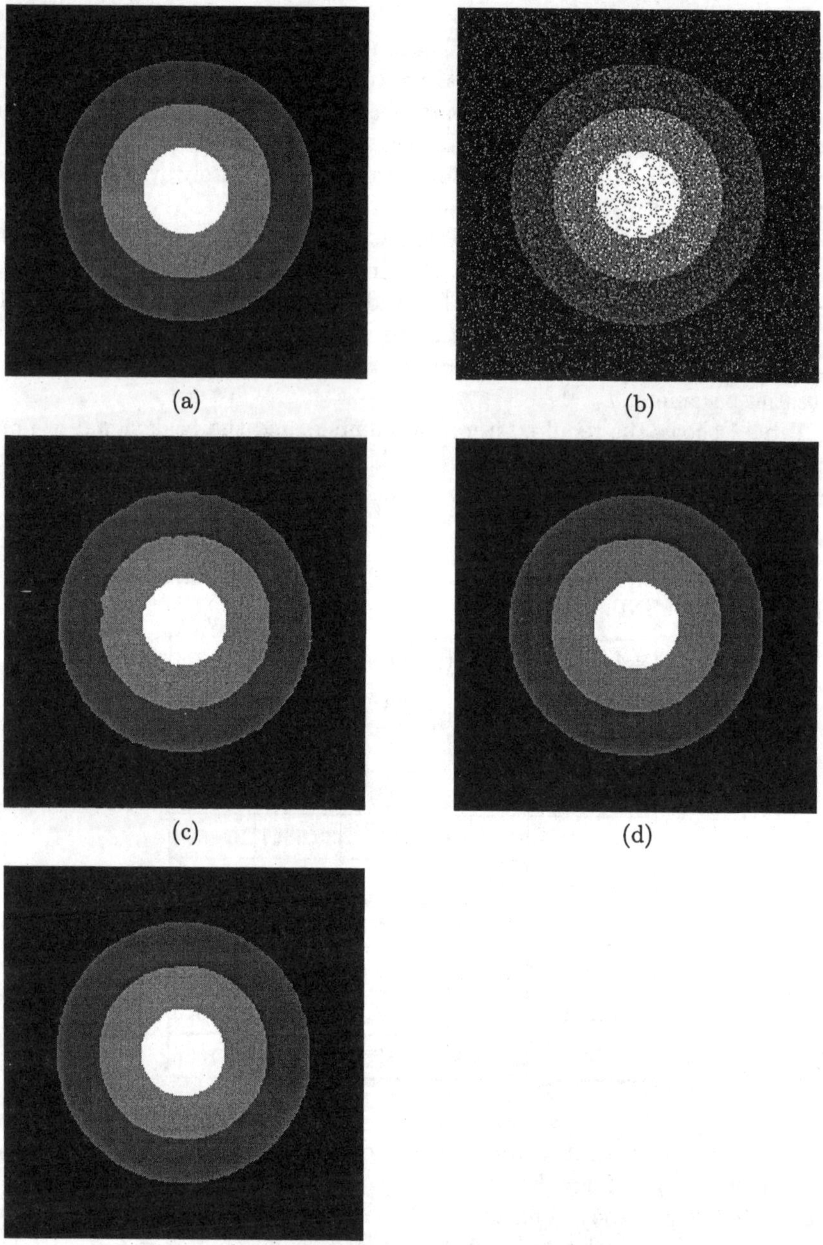

(a) (b) (c) (d) (e)

Fig. 2. Denoising of the circle image with noise level $\sigma = 0.5$: (a) Original image. (b) Noisy image. (c)-(e) Denoised images using the HNN, the TCNN and the NCNN, respectively

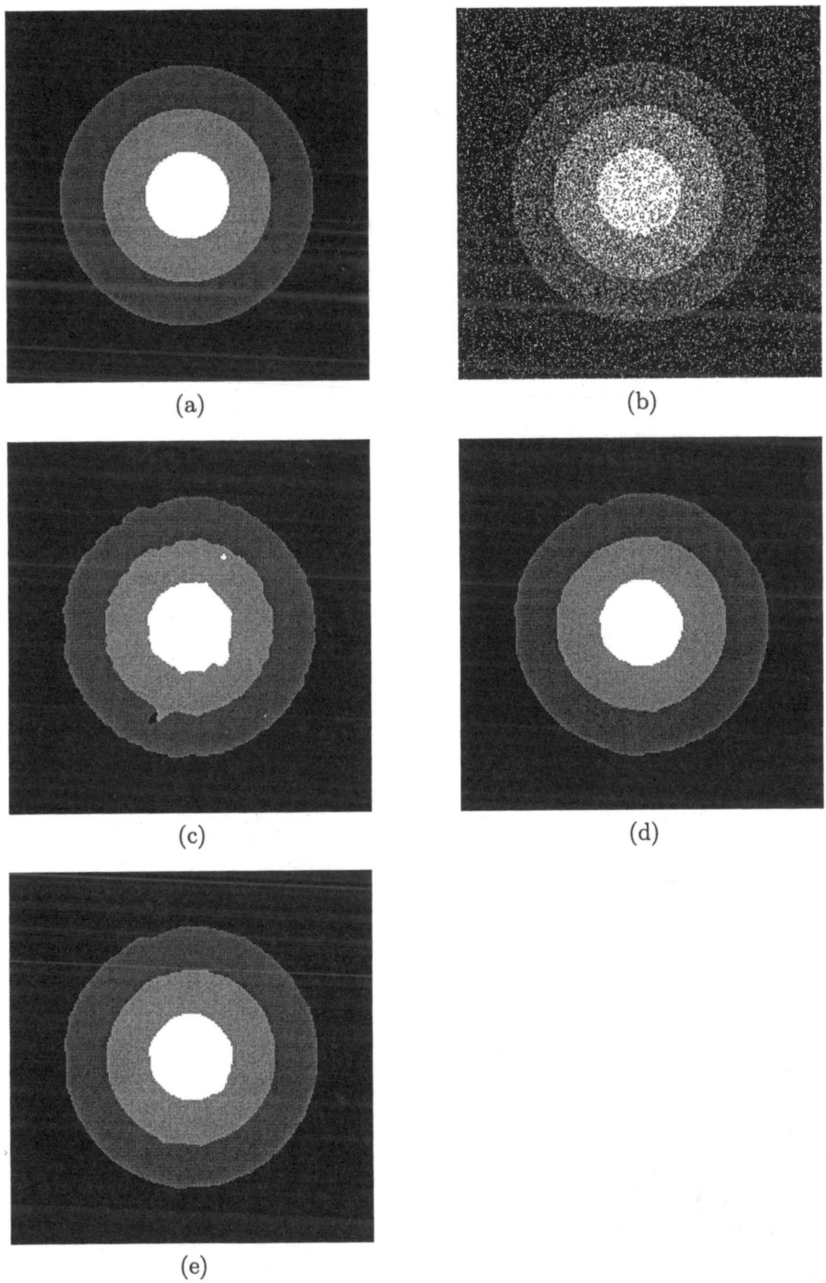

Fig. 3. Denoising of the circle image with noise level $\sigma = 0.75$: (a) Original image. (b) Noisy image. (c)-(e) Denoised images using the HNN, the TCNN and the NCNN, respectively

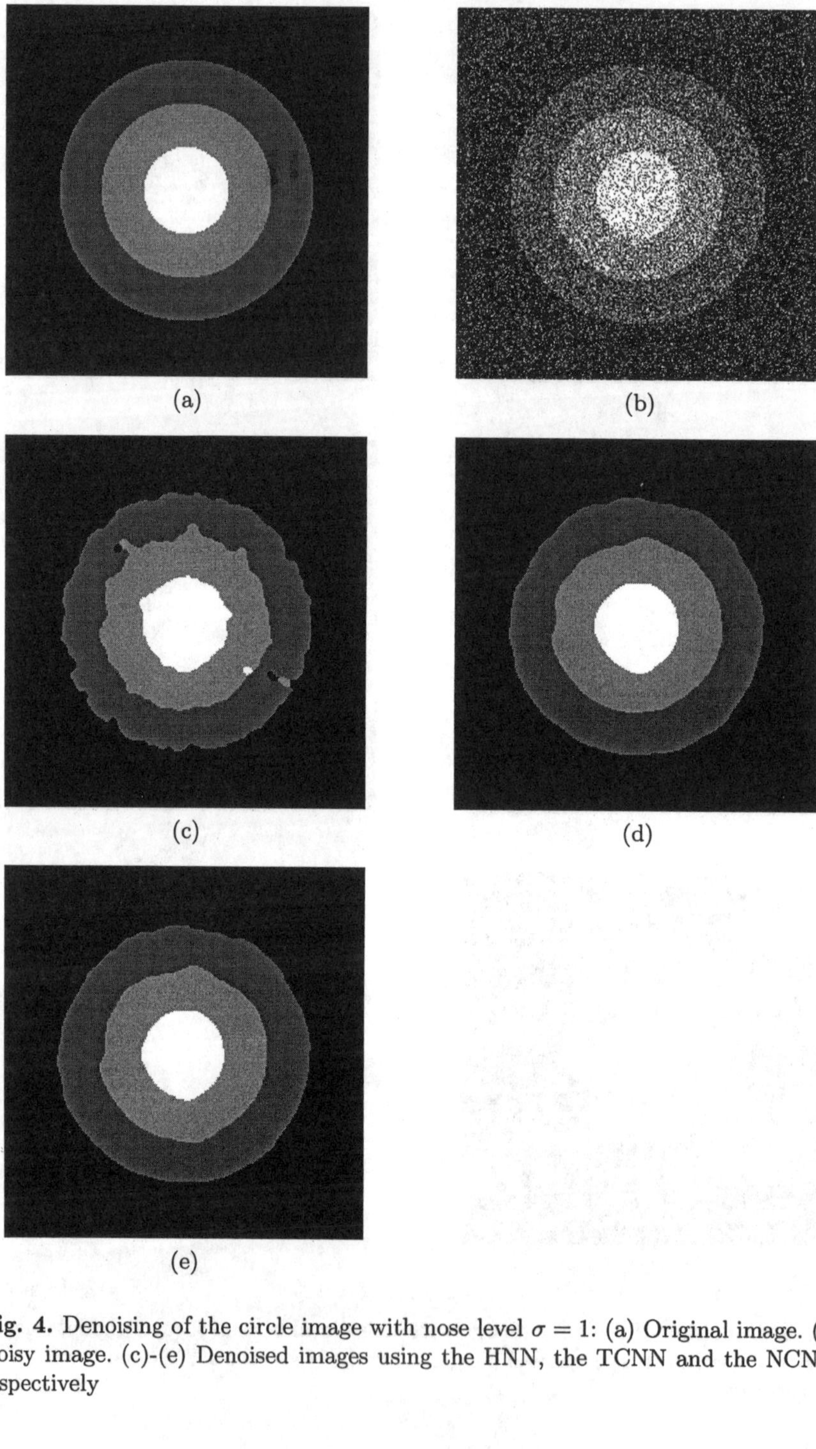

Fig. 4. Denoising of the circle image with nose level $\sigma = 1$: (a) Original image. (b) Noisy image. (c)-(e) Denoised images using the HNN, the TCNN and the NCNN, respectively

The decreasing rate of z and n, β_z and β_n are 0.005. For the TCNN and the HNN we use the same parameters as the NCNN except that $n = 0$ for the TCNN, $n = 0$ and $z = 0$ for the HNN. The MRF pair-site clique potential parameter β_c=8.

Table 2 shows the required iteration numbers and the peak signal-to-noise ratio (PSNR) of the denoised images. The higher the PSNR, the better the image quality. It can be seen from the table that the NCNN offers the best performance. The PSNR of the restored image using the NCNN is higher than those of the restored images using the HNN and the TCNN. In addition, the NCNN use less iterations to converge than the HNN and the TCNN. The denoised images are shown in Fig. 5-7.

Table 2. Numerical denoising results of the Lena image ($PSNR_r$: PSNR of the denoised image, $PSNR_d$: PSNR of the noisy image)

Noise Level		Iterations	$PSNR_r$	$PSNR_d$
$\sigma = 8$	HNN	950	31.2448	30.0532
	TCNN	835	31.5933	30.0532
	NCNN	807	32.0016	30.0532
$\sigma = 16$	HNN	1489	25.6227	23.0549
	TCNN	1173	25.9077	23.0549
	NCNN	1114	26.7554	23.0549
$\sigma = 24$	HNN	2054	22.8195	20.5695
	TCNN	1571	22.9979	20.5695
	NCNN	1862	23.1777	20.5695

5 Conclusion

A new neural network, called noisy chaotic neural network (NCNN), is used to address the MAP-MRF formulated image denoising problem. SCSA effectively overcomes the local minima problem. We have shown that the NCNN gives better quality solutions compared to the HNN and the TCNN.

References

1. Andrews, H. C., Hunt, B. R. (1977) Digital Image Restoration. Englewood Cliffs, N J, Prentice-Hall

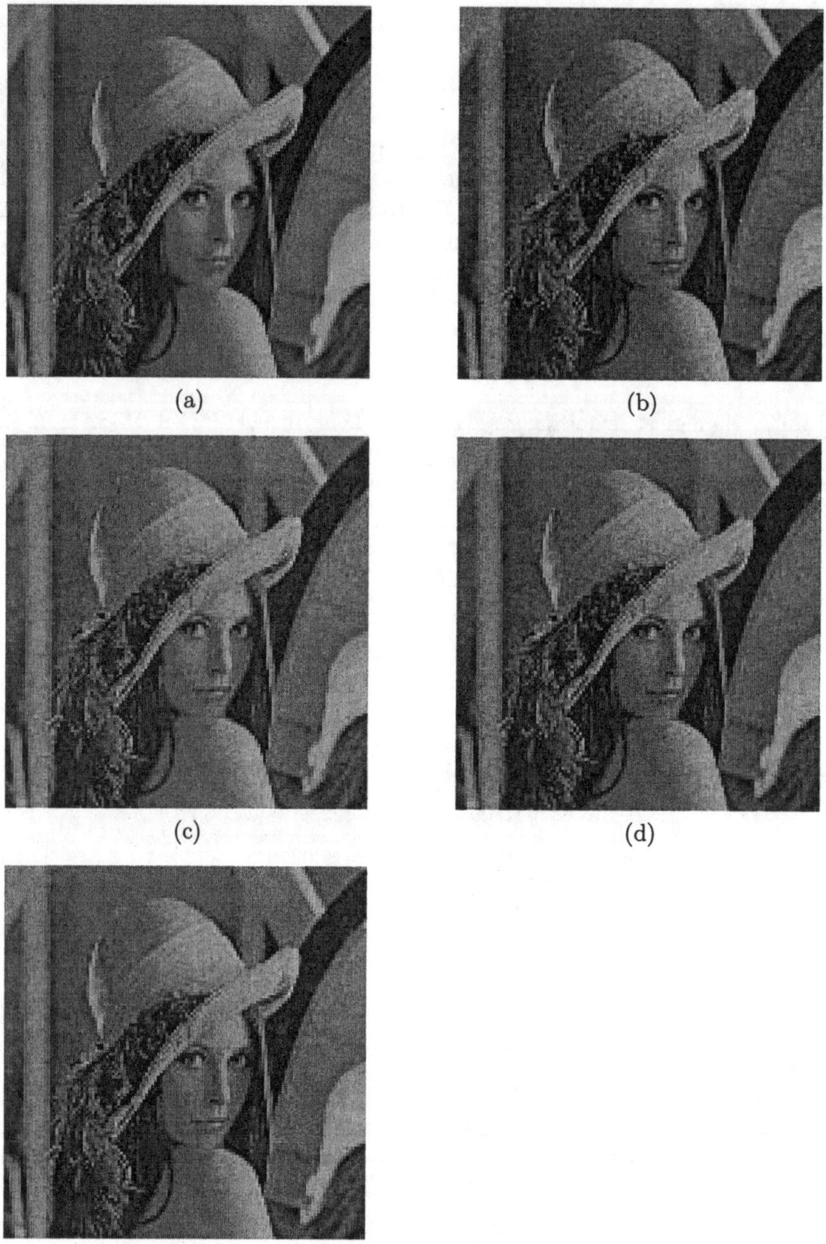

(a) (b) (c) (d) (e)

Fig. 5. Denoising of the Lena Image with noise level $\sigma = 8$: (a) Original image. (b) Noisy image. (c)-(e) Denoised images using the HNN, the TCNN and the NCNN, respectively

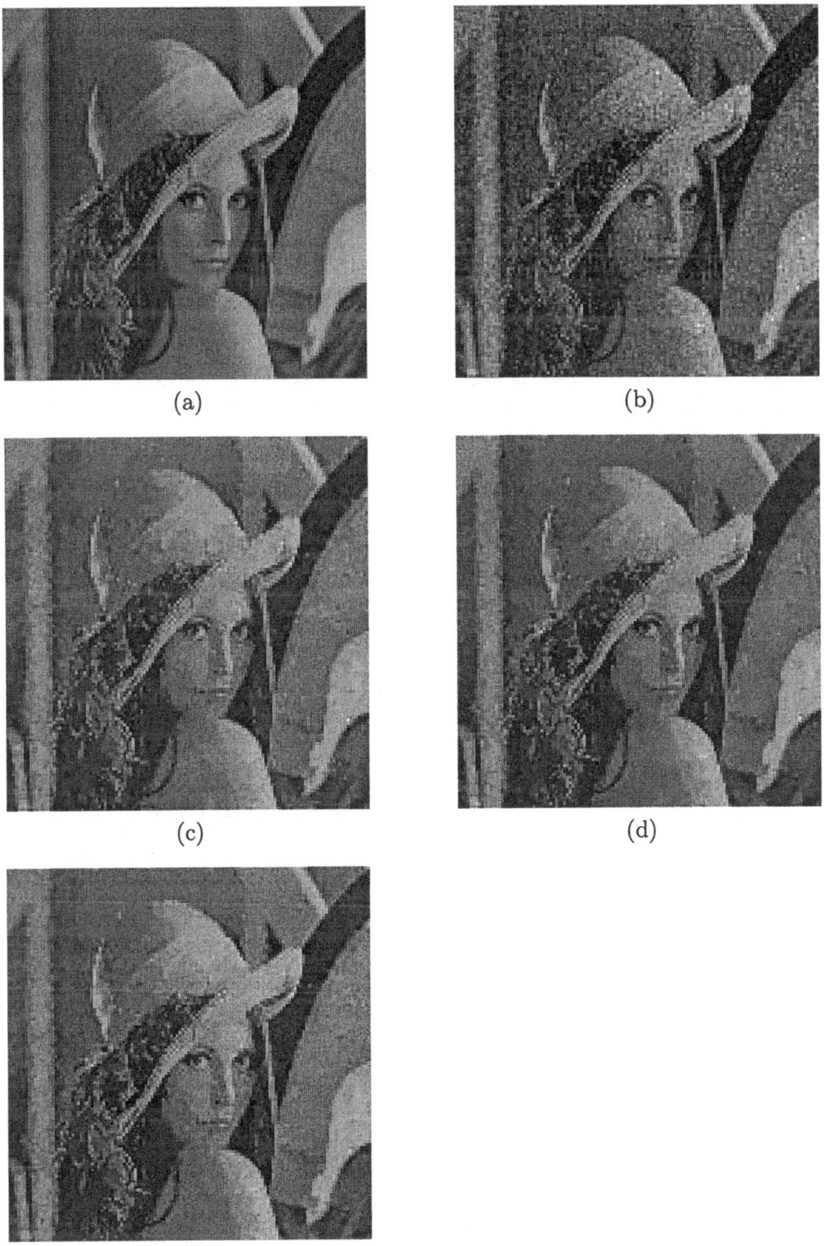

Fig. 6. Denoising of the Lena Image with noise level $\sigma = 16$: (a) Original image. (b) Noisy image. (c)-(e) Denoised images using the HNN, the TCNN and the NCNN, respectively

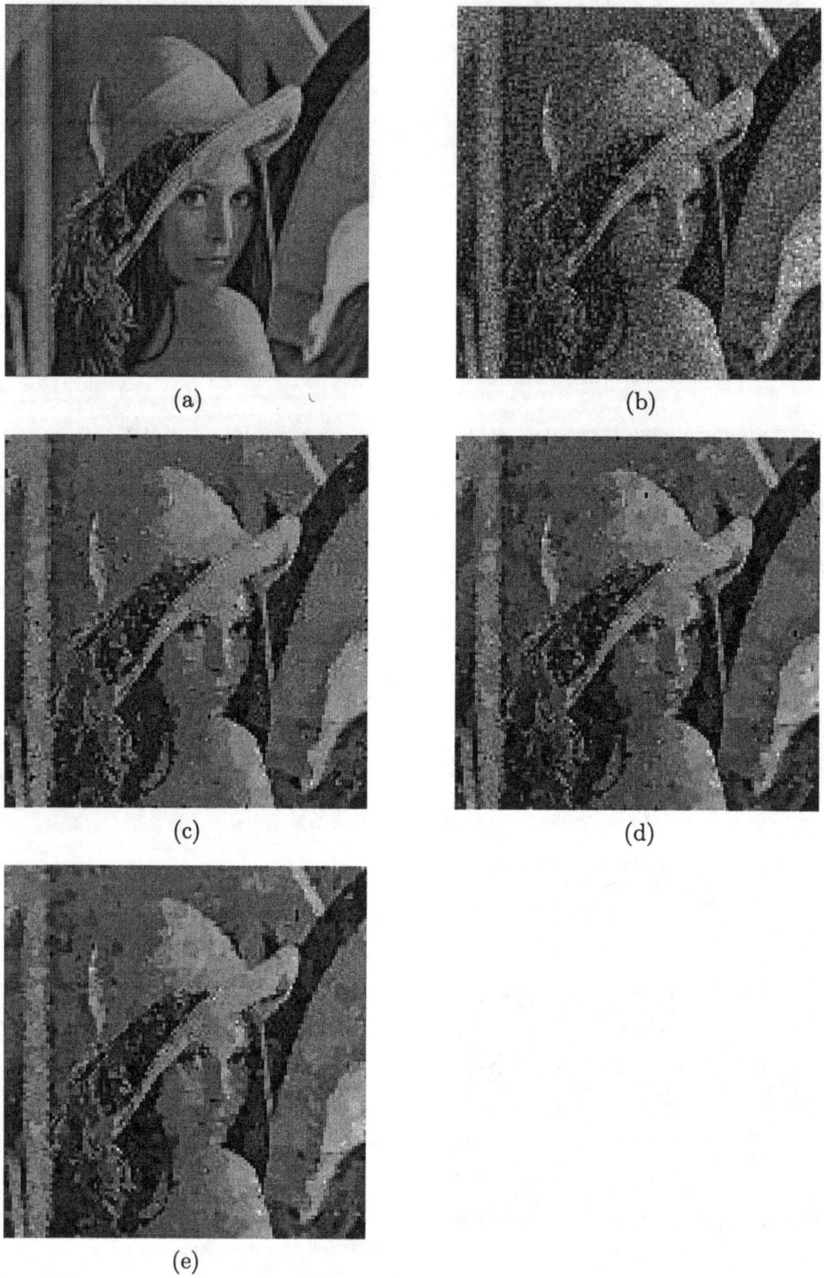

Fig. 7. Denoising of the Lena Image with noise level $\sigma = 24$: (a) Original image. (b) Noisy image. (c)-(e) Denoised images using the HNN, the TCNN and the NCNN, respectively

2. Rosenfeld, A., Kak, A. C. (1982) Digital Picture Processing, **1**, Academic Press, 2nd edition
3. Geman ,S., Geman, D. (1984) Stochastic relaxation, gibbs distributions and the bayesian restoration of images. IEEE Transactions on Pattern Analysis and Machine Intelligence, **66**, 721-741
4. Li, S. Z. (1998) Map image restoration and segmentation by constrained optimization. IEEE Transactions on Image Processing, **712**, 1730-1735
5. Wang, H., Li, S. Z., Petrou, M. (1990) Relaxation labeling of markov random fields. IEEE Transactions on Systems, Man and Cybernetics, **20**, 709-715
6. Peterson, C., Soderberg, B. (1992) Combinatorial Optimization with Neural Networks. Blackwell
7. Hopfield, J. J., Tank, D. W. (1985) Neural computation of decisions in optimization problems. Biol. Cybern., **52**, 141-152
8. Chen, L., Aihara, K. (1995) Chaotic simulated annealing by a neural network model with transient chaos. Neural Networks, **86**, 915-930
9. Chen, L., Aihara, K. (1994) Transient chaotic neural networks and chaotic simulated annealing. Towards the Harness of Chaos, 347-352
10. Aihara, K., Tokuda, I., Nagashima, T. (1998) Adaptive annealing for chaotic optimization. Phys. Rev. E, **58**, 5157-5160
11. Kajiura, M., Anzai, Y., Akiyama, Y., Yamashira, A., Aiso, H. (1991) The gaussian machine: a stochastic neural network for solving assignment problems. J. Neural Network Comput., **2**, 43-51
12. Wang, L., Tian, F. (2000) Noisy chaotic neural networks for solving combinatorial optimization problems. Proc. International Joint Conference on Neural Networks, **4**, 4037-4040
13. Hammersley, J. M., Clifford, P. (1971) Markov field on finite graphs and lattices. unpublished Manuscript
14. Li, S. Z. (1995) Markov Random Field Modeling in Computer Vision. Springer-Verlag

Soft Computation of Numerical Solutions to Differential Equations in EEG Analysis

Mingui Sun[1], Xiaopu Yan[2], and Robert J. Sclabassi[1]

[1] Departments of Neurosurgery, University of Pittsburgh, Pittsburgh, PA 15260
mrsun@neuronet.pitt.edu
[2] Compunetix, Inc., Monroeville, PA 15146 pyan@compunetix.com

Abstract. Computational localization and modeling of functional activity within the brain, based on multichannel electroencephalographic (EEG) data are important in basic and clinical neuroscience. One of the key problems in analyzing EEG data is to evaluate surface potentials of a theoretical volume conductor model in response to an internally located current dipole with known parameters. Traditionally, this evaluation has been performed by means of either finite boundary or finite element methods which are computationally demanding. This paper presents a soft computing approach using an artificial neural network (ANN). Off-line training is performed for the ANN to map the forward solutions of the spherical head model to those of a class of spheroidal head models. When the ANN is placed on-line and a set of potential values of the spherical model are presented at the input, the ANN generalizes the knowledge learned during the training phase and produces the potentials of the selected spheroidal model with a desired eccentricity. In this work we investigate theoretical aspects of this soft-computing approach and show that the numerical computation can be formulated as a machining learning problem and implemented by a supervised function approximation ANN. We also show that, for the case of the Poisson equation, the solution is unique and continuous with respect to boundary surfaces. Our experiments demonstrate that this soft-computing approach produces highly accurate results with only a small fraction of the computational cost required by the traditional methods.

Keywords: artificial neural network, efficient algorithm, electroencephalography, forward solution, partial differential equation

1 Background Of EEG

The quest for understanding the working of the brain is of paramount interest. Although significant amounts of knowledge have been obtained, there are still more questions than answers. The human brain, which weights between 1.2 and 1.8 kilograms, consists of approximately one trillion cells. Among them, 100 billion cells are neurons which are the main sources of intelligence, memory, and consciousness[1]. The complexity of the brain is due not only to this extremely large number of cells, but also to the diversity of cells and their interconnections forming an integrated object which is often hailed as the most complex system in the universe[1]. Despite this complexity and diversity, all neurons have, in common, the functional properties of integration, conduction, and transmission of nerve impulses. The neuron consists of three parts: 1) a dendritic branching through which input information is transferred to the cell, 2) a body (or soma) which serves to integrate this information, and 3) an axon, which is a "cable" transferring information to other neurons. Each neuron is in contact through its axon and dendrites with other neurons, so that each neuron is an interconnecting element in the network of the nervous system.

Compact groups of neurons, called nuclei, are anatomically identifiable within the central nervous system. Tracts of axons connecting these nuclei can be traced from region to region and it is to such relatively complex nucleus regions that the various functions of the nervous system are related. Although the electric activity of a single neuron cannot be observed from the scalp, synchronized activity involving thousands of neurons can. These neurons produce electric current flowing within the volume conductor of the head, producing brain function related potential waves which form the main components of the observed electroencephalogram (EEG).

The EEG recorded from the scalp in man typically has amplitude values from 10 to 100 μV and a frequency content from 0.5 to 40 Hz. Signals of 10 to 30 μV are considered low amplitude and potentials of 80 to 100 μV are considered high amplitude. The spectrum of the EEG is traditionally divided into four dominant frequency bands: δ-band (0.5-4 Hz), θ-band (4-8 Hz), α-band (8-13 Hz), and β-band (13-30 Hz). An alert person displays a low amplitude EEG of mixed frequencies in the 8 to 13 Hz range, while a relaxed person produces large amounts of sinusoidal waves, at a single frequency in the 8 to 13 Hz range, which are particularly prominent at the back of the head.

Traditional EEG consisted of only one or several channels of data which were mechanically plotted on paper. Recent developments in computer technology have changed the methods of collecting, storing and analyzing EEG data significantly[2]. EEG data can now be acquired from tens or hundreds of scalp electrodes stored on digital media, such as large-capacity hard discs or DVD discs, with low cost.

Traditional EEG analysis relies on visual inspection by physicians. As the computers double their computational speed in approximately a single year, more complex computational tasks can now be performed in real-time. Large numbers of traces of data can now be sampled and processed simultaneously, along with other input data, such as video and audio, which are synchronously recorded with EEG episodes.

2 Forward Problem of EEG

Computational modeling of multichannel EEG measured on the scalp, based on electrostatic theory, helps us to understand where and how the brain processes information. This modeling is also important in the detection of neurological disorders, such as brain tumors and epilepsy. Two major problems are present in this modeling. One is the *forward problem* which computes theoretical scalp potentials excited by a known current source within a volume conductor, and the other is the *inverse problem* which localizes an underlying source within a volume conductor based on scalp potential measurements[4]. This work focuses on the forward problem aiming at reducing the computational complexity in EEG modeling.

A volume conductor can be considered as a distributed, passive, and linear circuit in which the relationship between an internal current source and its electric potential (or voltage) response provides a solution to the forward problem. This case is similar to that of a lumped, resistive circuit where the current-voltage relationship is given by the Ohm's law. In the volume conduction case, the current-voltage relationship is given by a partial differential equation (PDE). In order to illustrate the fundamental concepts, we provide the following derivation under certain assumptions. For more in-depth treatment, see standard texts on electrostatic theory such as [3, 5, 6].

Let us assume that there exists a bipolar current source with a current density $\bar{J}$ (a vector quantity) within a volume conductor. For the EEG case, we assume that $\bar{J}$ is generated by a compact group of active neurons described previously. These neurons convert chemical energy to electrical energy in a physical action similar to that of a battery. In the literature, $\bar{J}$ is often called *impressed current density* or *primary current density*[3, 4], and is defined to be zero outside the region of the source. If we let the volume of this region approach zero, in the limit, a bipolar volumetric current source becomes a *current dipole* located at a single point and $\bar{J}$ becomes the *dipole moment density* (to be discussed further in Section 3.1). The dipole source model is widely utilized in biophysical study of the EEG[4].

Let the conductivity value of the volume conductor be σ. The impressed current produces an electric field $\bar{E}$ which defines another form of current, a *return current* or *secondary current*, with a density $\bar{J}_1$ [3, 4]. In contrast to $\bar{J}$, the return current $\bar{J}_1$ exists only in the space outside the region of the source and is related to the electric field by

$$\bar{J}_1 = \sigma \bar{E}. \tag{1}$$

We remark that the definition of $\bar{J}_1$ is necessary because, without it, $\bar{J}$ would not have a return path.

The total current density $\bar{J}_s$ is the sum of $\bar{J}$ and $\bar{J}_1$, or

$$\bar{J}_s = \bar{J} + \sigma \bar{E}. \tag{2}$$

Now, let us assume that the electric field is *quasistatic*, which implies that the dielectric components of the brain tissue are negligible. This assumption is acceptable for the EEG case because, as discussed in Section 1, the major spectral components in the EEG are generally less than 40 Hz which does not cause significant dielectric effects[3, 4]. As a result, the source-potential (or input-output) relationship of the system is *instantaneous*. Then

$$\bar{E} = -\nabla\phi \tag{3}$$

where ϕ is the electric potential (a scalar) which, in our case, represents the EEG. Since the flow lines of $\bar{J}_s$ within the volume conductor are closed, e.g., $\bar{J}_s$ is *solenoidal* in terms of the vector field theory[7], the divergence of $\bar{J}_s$ is zero. Thus, by (2) and (3)

$$\nabla\bar{J}_s = \nabla\bar{J} - \nabla\cdot\sigma\nabla\phi = 0. \tag{4}$$

If a homogeneous medium is assumed, we have

$$\sigma\nabla^2\phi = \nabla\cdot\bar{J}. \tag{5}$$

Equation (5) is the well-known Poisson equation which plays a central role in the forward problem of the EEG.

3 Forward Solutions

For volume conductors of arbitrary shapes, the PDEs in (4) or (5) generally does not have an analytic solution. However, for volume conductors of simple shapes, analytic solutions can be found. In the following, we present several special cases without detailed derivations. In all these cases, we assume that the current source is an ideal single dipole located at position $\mathbf{r}$ with current moment vector $\mathbf{m}$. These special cases have been applied to EEG modeling and are useful for soft-computation of scalp potentials based on the ANN approach.

3.1 Infinite Homogeneous Medium

Theoretically, the potential response to a single dipole can be considered as a sum of potential responses to two monopoles which generate currents with the same amplitude but opposite polarities. This theoretical treatment is convenient because: (1) there exists an important analogy between the potential field produced by a current monopole and the electrostatic field produced by a point charge, and (2) the linearity of the system validates the use of the superposition principle. For a current monopole located at $\mathbf{r}'$ emitting current I_0, the monopolar potential, $\phi_m(\mathbf{r})$, can be written using the analogy to the electrostatic field[5]:

$$\phi_m(\mathbf{r}) = \frac{I_0}{4\pi\sigma|\mathbf{r}-\mathbf{r}'|}. \tag{6}$$

Now, let us place a "movable" monopole of current $-I_0$ in the vicinity of the first monopole at a distance δ, set $\delta I_0 \bar{\mathbf{d}}_0 = \mathbf{m}$ with $\bar{\mathbf{d}}_0$ being a unit vector specifying the line direction connecting the two monopoles, and take a limit. As $\delta \longrightarrow 0$, the two monopoles overlap resulting in an ideal dipole. With this approach, it is simple to show[5] that the potential $\phi(\mathbf{r})$ for a single dipole in an infinite homogeneous medium is given by

$$\phi(\mathbf{r}) = \frac{\mathbf{m}}{4\pi\sigma} \cdot \frac{\mathbf{r}-\mathbf{r}'}{|\mathbf{r}-\mathbf{r}'|^3}. \tag{7}$$

3.2 Homogeneous Sphere

The analytic solution, $\phi_u(\mathbf{r})$, of Poisson equation for the unit-radius spherical volume conductor with $|\mathbf{r}| = 1$ and a homogeneous conductivity has the following simple form[8]:

$$\phi_u(\mathbf{r}) = \frac{1}{4\pi\sigma q} \sum_{i=1}^{3} m_i \left[\frac{2(x_i - x_i')}{q^2} + e_i + \frac{x_i s - x_i'}{q+1-t} \right] \tag{8}$$

where the center of the unit sphere is at the origin; x_i, x_i' and m_i, $i = 1, 2, 3$, are, respectively, the elements of vectors $\mathbf{r}'$, $\mathbf{r}$, and $\mathbf{m}$ representing the dipole location, scalp location, and current moment; and q and t are equal to, respectively, $|\mathbf{r}-\mathbf{r}'|$ and $\mathbf{r}\cdot\mathbf{r}'$.

For a spherical volume conductor with an arbitrary radius R, the surface potential, $\phi(\mathbf{r})$, for a single dipole in a homogeneous sphere is given by

$$\phi(\mathbf{r}) = \frac{\phi_u(\mathbf{r})}{R^2}. \tag{9}$$

3.3 Multishell Sphere

The head can be modeled as a sphere of radius R centered at the origin having several concentric-shells of volume conductors. Fig. 1 shows a special case with four-shells representing the brain, skull, cerebrospinal fluid, and scalp. The forward solution relating the dipole located at $\mathbf{r}'$ to the potential value $\phi(\mathbf{r})$ at the surface point $\mathbf{r}$ is given by[9, 10]

$$v = \frac{1}{4\pi\sigma_4 R^2} \sum_{n=1}^{\infty} c_n f^{n-1} \mathbf{m} \cdot \left[\mathbf{r}'_0 P_n(\cos\theta) + \mathbf{t}_0 \frac{P_n^1(\cos\theta)}{n} \right] \tag{10}$$

where the dot after $\mathbf{m}$ denotes dot product; σ_4 is the conductivity value for the scalp (see Fig. 1); $\mathbf{r}'_0$ and $\mathbf{t}_0$ are the radial and tangential unit vectors,

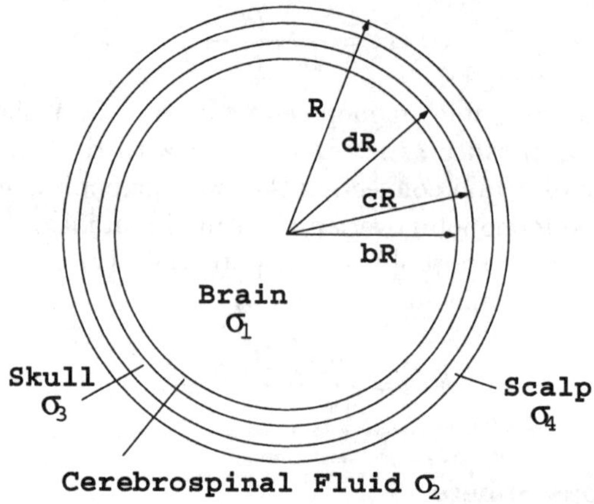

Fig. 1. Central cross-section of the 4-shell spherical volume conductor model. The brain, cerebrospinal fluid, skull and scalp regions are illustrated, and their respective conductivity values are labeled by $\sigma_i, i = 1, 2, 3, 4$. The boundaries of these regions are located at bR, cR, and dR, where R is the radius of the head model.

respectively; $f = |\mathbf{r}'|/R$ represents the eccentricity of the dipole; c_n denotes a series determined by the model geometry and conductivity values; θ denotes the angle between **r'** and **r**; and $P_n(\cos\theta)$ and $P_n^1(\cos\theta)$ are, respectively, the Legendre and associated Legendre polynomials of order n.

The unit vector $\mathbf{t}_0$ in (10) is given in cross products[9]:

$$\mathbf{t}_0 = \frac{\mathbf{t}}{|\mathbf{t}|} \tag{11}$$

with

$$\mathbf{t} = \mathbf{r}' \times \mathbf{r} \times \mathbf{r}' = \mathbf{r}(\mathbf{r} \cdot \mathbf{r}') - \mathbf{r}'(\mathbf{r} \cdot \mathbf{r}). \tag{12}$$

Assuming that the head model has four concentric shells as shown in Fig. 1, the expression for c_n is given by[11]:

$$c_n = \frac{(2n+1)^4 (cd)^{2n+1}}{\Gamma} \tag{13}$$

with

$$\begin{aligned}\Gamma = d^{2n+1}&\Big[b^{2n+1}n(\frac{\sigma_1}{\sigma_2}-1)(\frac{\sigma_2}{\sigma_3}-1)(n+1)\\
&+c^{2n+1}(\frac{\sigma_1}{\sigma_2}n+n+1)(\frac{\sigma_2}{\sigma_3}n+n+1)\Big]\\
&\times\Big[(\frac{\sigma_3}{\sigma_4}n+n+1)+(n+1)(\frac{\sigma_3}{\sigma_4}-1)d^{2n+1}\Big]\\
&+(n+1)c^{2n+1}\Big[b^{2n+1}(\frac{\sigma_1}{\sigma_2}-1)(\frac{\sigma_2}{\sigma_3}n+\frac{\sigma_2}{\sigma_3}+n)\\
&+c^{2n+1}(\frac{\sigma_1}{\sigma_2}n+n+1)(\frac{\sigma_2}{\sigma_3}-1)\Big]\\
&\times\Big[n(\frac{\sigma_3}{\sigma_4}-1)+(\frac{\sigma_3}{\sigma_4}n+\frac{\sigma_3}{\sigma_4}+n)d^{2n+1}\Big]\end{aligned} \tag{14}$$

where $b, c, d < 1$, respectively, denote the outer-most radii (relative to the radius of the sphere R) for the brain (conductivity value $\sigma = \sigma_1$), cerebrospinal fluid ($\sigma = \sigma_2$), and skull ($\sigma = \sigma_3$). The values for a, b, c and conductivity constants for different head models have been tabulated in [12].

In practice, it is impossible to compute the forward solution in (10) exactly. The approximation of this solution is usually performed by summing 60 to 100 terms indexed by n[8]. In order to facilitate computation, we have developed an efficient algorithm implemented by a short C-program[10, 13], which provides drastic savings in computational cost. The required floating-point operations have been reduced from over 1,500 to about 100.

3.4 Homogeneous and Multishell Spheroids

A spheroid is a special case of an ellipsoid in which two of the three axes are equal. The analytical solution to Poisson equation for the homogeneous spheroidal volume conductor has been reported by Yeh and Martinek[14]. Since the mathematical expressions for this solution are extremely complex, we will not include them here. Instead, we briefly state the computational procedures: 1) transform the Cartesian coordinates to the prolate spheroidal coordinates; 2) evaluate the Hessian matrix of the transformation; 3) compute three types of associated Legendre functions of order 0 to "infinity"; and 4) compute nested infinite sums with respect to two indices of the associated Legendre functions and their derivatives.

As in the case of the spherical model, the interior of the spheroidal model can be divided into a number of shells with different conductivity values. Although the analytic form of the forward solution exists for the multishell spheroidal model[15], the mathematical expressions are even more complex than the homogeneous case. Again, we do not include them here. Due to the high complexity and numerical instability involved, the analytic expression for the multishell spheroidal model is not widely utilized.

3.5 Finite and Boundary Element Models

Although the volume conductor models presented previously have analytic forward solutions, their accuracy is limited due to the shape difference between the model and the head[16]. More precise models utilize the finite boundary element method (BEM) or finite element method (FEM) built upon realistic internal and external structures of the head; however, the computational costs of these models are usually several orders of magnitude higher than that of the spherical model. Although faster computers and continuous improvement of the BEM and FEM algorithms have accelerated computation, currently the computational speed is still unsatisfactory.

4 Soft-Computing Approach

We have developed a novel approach to solving the Poisson equation using non-spherical volume conduction models by soft-computing[17, 18, 19]. Instead of decomposing the object shape into numerous 2-D or 3-D elements and "hard-computing" the resulting matrices of extremely large sizes, we parameterize the shape and compute the desired solution according to a known solution to a standard shape using a pre-trained artificial neural network. The following procedure is utilized: 1) The input shape is analyzed to determine the optimal shape parameters. This process is exemplified in Fig. 2 for the case of a human head where the standard shape is a sphere (dotted curve in the left panel and circle in the right panel). Two parameters (radius and eccentricity) are determined to define a spheroid (dash curve in the left panel) which best-fits the head shape. 2) Both the shape parameters and the solution values for the standard shape are utilized, after appropriate scaling, to form a pattern vector as the input to a backpropagation (BP) artificial neural network (ANN). 3) The ANN extrapolates the solutions to a variety of shapes with different parameters that the ANN has experienced during an off-line training process. The PDE solution corresponding to the input shape is produced at the ANN output.

When compared to direct numerical evaluation using BEM or FEM, the ANN approach does not require volumetric meshing and equation solving. The elimination of these two expensive procedures greatly reduces the on-line computational cost, allowing real-time PDE solving on ordinary personal computers. Although heavy computation is required to train the network when the computational system is constructed, this is a one-time procedure which does not affect the on-line performance.

5 Theoretical Considerations

In this section we answer three fundamental questions: First, whether the ANN is capable of approximating the complex functional relationship between

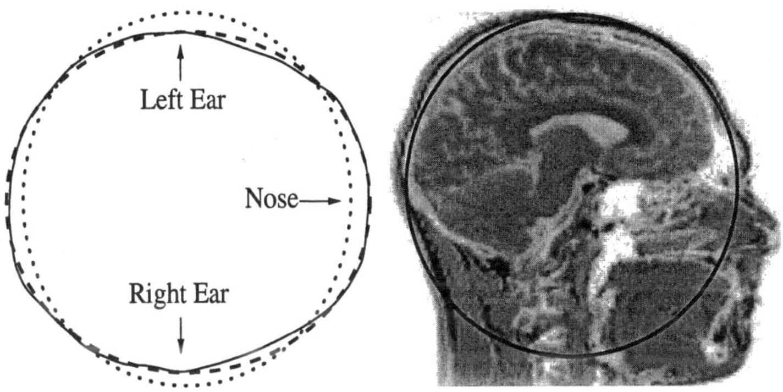

Fig. 2. Comparison between Head and Model Shapes. Left: Cross-section view of a head shape (solid curve) fit by a sphere (dot curve) and a spheroid (dash curve). Right: Side view of an MRI image with the best fit of the head shape. *(Reused figure with permission from [17] @ 2000 IEEE).*

solutions with different boundary surfaces? Second, is the solution unique under a fixed boundary surface? Third, whether the solution is continuous with respect to the boundary surface? If any of these three questions has a negative answer, the ANN method will not behave appropriately. For instance, the computed result will be ambiguous if the solution is non-unique, and will be unstable if the solution with respect to the boundary condition is not continuous.

5.1 Function Approximation

The answer to the first question is clear. The universal approximation theorem[20] indicates that a class of ANNs, such as the nonlinearly activated backpropagation, radial basis, and generalized regression ANNs can approximate any continuous function to an arbitrarily small error provided that the numbers of training samples and hidden units are sufficiently large. Therefore, the explicit form of the functional relationship between PDE solutions is not required in the ANN approach. A highly systematic training procedure can be utilized to teach the knowledge-based system. The learned relationships are stored in the associative memory of the system and re-assembled to produce an output when an inquiry is present at the input.

5.2 Uniqueness of Solution

We now present the uniqueness of the forward solution provided by Poisson PDE $\nabla^2\phi = \frac{1}{\sigma}\nabla \cdot \bar{J}$, where $\nabla^2\phi = \frac{\partial^2\phi}{\partial x^2} + \frac{\partial^2\phi}{\partial y^2} + \frac{\partial^2\phi}{\partial z^2}$ and $\nabla \cdot \bar{J}$ specifies the source density. In electrostatics the uniqueness problem has been studied with respect to two types of boundary conditions[6]. The *Dirichlet boundary condition* specifies the potential on a closed boundary surface, denoted by S, while the *Neumann boundary condition* specifies the derivatives at the normal direction of the same surface. To show the uniqueness, we contrarily assume that there exist two solutions ϕ_a and ϕ_b that satisfy the same boundary conditions on S. Let us define $u = \phi_a - \phi_b$. We have

$$\nabla^2 u = \nabla^2 \phi_a - \nabla^2 \phi_b = \frac{1}{\sigma}\nabla \cdot \bar{J} - \frac{1}{\sigma}\nabla \cdot \bar{J} = 0 \tag{15}$$

which is true within the entire volume V inside S. Green's first theorem[21] states:

$$\int_V (p\nabla^2 q + \nabla p \cdot \nabla q) dv = \oint_S p \frac{\partial q}{\partial n} ds \tag{16}$$

where p, q, and their Laplacians are defined in V. Setting $u = p = q$, and use (15), we have

$$\int_V |\nabla u|^2 dv = \oint_S u \frac{\partial u}{\partial n} ds. \tag{17}$$

Since, by definition, ϕ_a and ϕ_b satisfy the same boundary conditions, either $u = 0$ (for the Dirichlet boundary condition) or $\frac{\partial u}{\partial n} = 0$ (for the Neumann boundary condition) holds true. Therefore, the right side of (17) is zero, which implies $\nabla u = 0$. Consequently, u must be a constant. For the Dirichlet case, $\phi_a = \phi_b$ in V since $u = 0$ on S. For the Neumann case, on the other hand, the solution is unique, differing only by a constant. In bio-potential problems the Dirichlet and Neumann conditions are often jointly applied. An extension of the above results indicate that the solution is unique.

5.3 Continuity of Solution

To answer the third question, we must show that, if the variation of the boundary condition is small, the variation of the solution is also small. This continuity condition has fundamental importance in the success of the ANN approach. We have investigated this condition and shown that it is indeed satisfied for the Poisson case.

Without loss of generality, we consider a Poisson PDE with the Dirichlet boundary condition:

$$\begin{aligned} &\Delta^2 \phi = \frac{1}{\sigma}\Delta \cdot \bar{J}, \quad in \;\; V_1 \\ &\phi = g_1 \quad on \;\; \partial V_1 \end{aligned} \tag{18}$$

$$\begin{aligned} &\Delta^2 \phi = \frac{1}{\sigma}\Delta \cdot \bar{J}, \quad in \;\; V_2 \\ &\phi = g_2 \quad on \;\; \partial V_2, \end{aligned} \tag{19}$$

where $\partial V_i = S_i$, $i = 1, 2$, denotes closed boundaries S_i of domains V_i. These two domains are very close to each other, i.e., for a small δ, for any point p_1 on $\partial V_1 \cap V_2$ (or $\partial V_2 \cap V_1$) there exists $p_2 \in \partial V_2$ (or $p_2 \in \partial V_1$) such that

$$|p_1 - p_2| \leq \delta. \tag{20}$$

And g_1 and g_2 are assumed to be very close on $\partial V_1 \cap \partial V_2$, i.e. for a small $\epsilon > 0$,

$$|g_1(p) - g_2(p)| < \epsilon \quad \text{for} \quad p \in \partial V_1 \cap \partial V_2. \tag{21}$$

In our application, V_1 and V_2 can be considered as, respectively, a simplified shape model, such as a spherical model, and a real head shape.

Let V be a larger domain containing both V_1 and V_2, and let $\tilde{\phi}_1$ and $\tilde{\phi}_2$ be extensions, defined on V, of ϕ_1 and ϕ_2

$$\begin{array}{ll} \tilde{\phi}_1 = \phi_1 & in \;\; V_1 \\ \tilde{\phi}_1 = g_1 & in \;\; V - V_1 \end{array} \tag{22}$$

$$\begin{array}{ll} \tilde{\phi}_2 = \phi_2 & in \;\; V_2 \\ \tilde{\phi}_2 = g_2 & in \;\; V - V_2 \end{array} \tag{23}$$

where ϕ_1 and ϕ_2 are solutions to (18) and (19), respectively.

It has been shown[22] that the following "well-poseness" estimate exists for solution ϕ:

$$||\phi||_{H^1(V)} \leq C(V)\big(||\nabla \cdot \bar{J}||_{L^2(V)} + ||g||_{H^{\frac{1}{2}}(\partial V)}\big) \tag{24}$$

where $||\cdot||_{H^1(\cdot)}$ and $||\cdot||_{H^{\frac{1}{2}}(\cdot)}$ are Sobolev norms of orders 1 and $\frac{1}{2}$, respectively. The details of this estimate have been provided in the literature[22].

Let $\tilde{\phi}_1$ and $\tilde{\phi}_2$ be two solutions in V and define

$$\tilde{\phi}_1 = \begin{cases} \phi_1 & \text{in } V_1 \\ \phi = g_1 & \text{in } V - V_1 \end{cases} \tag{25}$$

and

$$\tilde{\phi}_2 = \begin{cases} \phi_2 & \text{in } V_2 \\ \phi = g_2 & \text{in } V - V_2 \end{cases} \tag{26}$$

Let $\phi = \tilde{\phi}_1 - \tilde{\phi}_2$ in V. Then, ϕ satisfies

$$\begin{cases} \nabla^2 \phi = 0 & \text{in } V_1 \cap V_2 \\ \phi = \tilde{\phi}_1 - \tilde{\phi}_2 & \text{on } \partial(V_1 \cap V_2) \end{cases} \tag{27}$$

Since

$$\partial(V_1 \cap V_2) = (\partial V_1 \cap V_2) \cup (V_1 \cap \partial V_2) \cup (\partial V_1 \cap \partial V_2)$$

We have

$$\phi_{\partial(V_1 \cap V_2)} = \begin{cases} g_1 - \phi_2 & \text{on } \partial V_1 \cap V_2 \\ \phi_1 - g_2 & \text{on } V_1 \cap \partial V_2 \\ g_1 - g_2 & \text{on } \partial V_1 \cap \partial V_2. \end{cases} \tag{28}$$

When V_1 is very close to V_2, for any point p_1 on $\partial V_1 \cap V_2$, there exists $p_2 \in \partial V_2$, such that

$$|p_1 - p_2| \leq \nabla \qquad \text{and}$$

$$|\phi_2(p_1) - \phi_2(p_2)| < \epsilon. \tag{29}$$

Thus, on $\partial V_1 \cap V_2$, we have

$$\begin{aligned} |g_1(p_1) - \phi_2(p_1)| &= |g_1(p_1) - g_2(p_2) + \phi_2(p_2) - \phi_2(p_1)| \\ &= |g_1(p_1) - g_1(p_2) + g_1(p_2) \\ &\quad - g_2(p_2) + \phi_2(p_2) - \phi_2(p_1)| < 3\epsilon. \end{aligned} \tag{30}$$

Similarly, we can obtain that on $V_1 \cap \partial V_2$

$$|\phi_1 - g_2| < 3\epsilon. \tag{31}$$

On the other hand, from the assumptions we can easily have

$$|g_1 - g_2| < \epsilon \qquad \text{on } \partial V_1 \cap \partial V_2. \tag{32}$$

From (30), (31), and (32), we have

$$|\phi_{\partial(V_1 \cap V_2)}| < 3\epsilon, \quad \text{or} \quad ||\phi||_{H_{\frac{1}{2}}(\partial(V_1 \cap V_2))} \leq C \cdot \epsilon \tag{33}$$

which leads to, by (24)

$$||\phi_1 - \phi_2||_{H^1(V_1 \cap V_2)} \leq C'(V_1 \cap V_2)\epsilon. \tag{34}$$

This estimation implies that the solution ϕ_1 for the Dirichlet boundary problem on domain V_1 can be used to approximate the solution ϕ_2 for the problem on the domain V_2 as long as V_1 is close to V_2 and the boundary $\partial(V_\cap V_2)$ is smooth.

6 Experimental Design

6.1 Model Selection

Although we have demonstrated that the ANN approach has a solid theoretical foundation, its full-scale implementation requires a considerable effort for data pre-processing. For example, in order to map between the multishell spherical and realistic head models, the following major pre-processing steps are required: 1) image segmentation, registration, and reconstruction based on MRI or CT head scans; 2) surface or volumetric mesh building in 3D; and 3) implementation of BEM or FEM to obtain target training samples. In order to by-pass these procedures and concentrate this feasibility study on the ANN approach, our investigation focuses on a simple case where the forward solutions of a unit-radius homogeneous sphere are mapped to that of prolate spheroids of different eccentricities. Since both the spherical and spheroidal models have closed-form analytical solutions, we do not have to rely on BEM or FEM to obtain forward solutions. As a result, the ANN design is not affected by the numerical errors resulting from BEM or FEM. Using this simple case we expect to: 1) test the feasibility of the ANN approach, and 2) study the ANN structure and training techniques.

6.2 Model Design

The human head is close to a spheroid since the anterior-posterior axis (denoted by a) is longer than the lateral axis (denoted by b). The left panel in Fig. 2 shows a top view of a head shape (solid curve), the best-fit sphere (dotted curve), and the best-fit spheroid (dash curve). It is clear that the spheroid model provides a considerably better fit than the spherical model. The value of eccentricity η (defined by $\sqrt{1-\eta^2} = \frac{b}{a}$) varies among individuals. We estimated that η is between 0.4 and 0.6 for the general population. For the spheroidal model eccentricity η is the only parameter related to its shape. We also assume, without loss of generality, $b = 1$, since the surface potentials on the unit spheroid can always be scaled to fit different head sizes. Based on these assumptions, the head space, Ω, is given by

$$\frac{x^2+y^2}{1-\eta^2} + z^2 = a^2, \qquad 0.4 < \eta \le 0.6. \tag{35}$$

In the spherical base model we assume that its radius is equal to one. For analytical convenience we let the two models share the same coordinate system centered at origin o as illustrated in Fig. 3. The source-potential relationships for both the spherical and spheroidal models have been presented in Section 3. Although the spherical model is easy to implement, the spheroidal model requires evaluations of both the functions and their derivatives of three types of associated Legendre functions for orders from zero to infinity. These functions and derivatives can be expressed using various recursive relations; however, many of these recursions are numerically unstable. This problem has been addressed in [23] where stable algorithms have been presented. We utilized 30 terms to approximate each infinite sum. The accuracy in Legendre function evaluation was checked against the tabulated values in [23]. The entire accuracy of potential computation was examined by setting η close to zero and comparing the result to that obtained from (8).

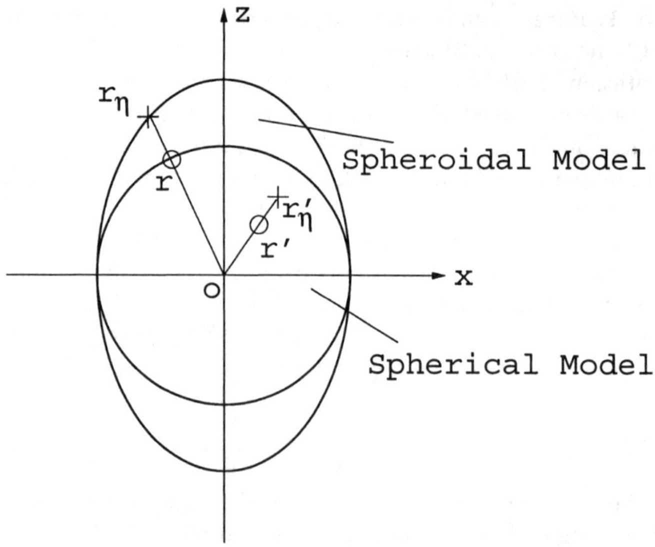

Fig. 3. One-to-one mappings between the electrode sites ($\mathbf{r}$ and $\mathbf{r}_\eta$) and between dipole locations ($\mathbf{r}$' and $\mathbf{r}_\eta$) for the spherical model and the spheroidal model.

6.3 Generation of Training Patterns

In order for the ANN to approximate the functional mapping between forward solutions of different models, we utilized training patterns constructed from densely located unit-strength random dipoles. For each training pattern, we first generated a large set of 3-element vectors for dipole location $\mathbf{r}'$ using the uniform probability distribution in the range $[-0.84, 0.84]$ which represents the boundary between the brain and the skull. Then, we discarded all vectors in this set whose modulus $|\mathbf{r}'|$ was greater than 0.84. The remaining 12,000 dipoles densely covered the brain region ($|\mathbf{r}'| < 0.84$) within the spherical model.

We associated with each dipole a current moment vector $\mathbf{m}$. Each of the three elements in $\mathbf{m}$ was first generated using the zero-mean, unit-variance Gaussian distribution. Then, $\mathbf{m}$ was normalized by re-signing it with $\mathbf{m}/|\mathbf{m}|$. Next, each ($\mathbf{r}$',$\mathbf{m}$) pair was plugged into Eq. (8) to compute a 20-element vector $\boldsymbol{\phi}_1$ at scalp locations defined by the International 10-20 system[24].

For each $\boldsymbol{\phi}_1$ of the spherical model we matched it with a 20-element vector $\boldsymbol{\phi}_2$ of the spheroidal model using the following procedure: 1) generate a random number for eccentricity η using the uniform probability distribution in the range between 0.4 and 0.6; 2) define a one-to-one mapping between the scalp electrode sites of the two models as shown in Fig. 3 where a ray is projected from the center at $\mathbf{o}$, through the electrode site at $\mathbf{r}$ on the spherical model, to the electrode site $\mathbf{r}_\eta$ on the spheroidal model (see the Appendix for details of calculation); 3) define a one-to-one mapping (again see Fig. 3 and the Appendix) between dipole locations by projecting from c through $\mathbf{r}'$ to $\mathbf{r}'_\eta$. and, 4) plug (η, $\mathbf{r}'_\eta$, $\mathbf{m}$, $\mathbf{r}_\eta$) into the analytical form of the spheroidal model to compute the 20-element potential vector ϕ_2.

6.4 Structure of ANN

A backpropagation (BP) ANN was utilized to approximate f in (15). In our ANN design (see Fig. 4) we utilized a single hidden layer with a bipolar sigmoid activation function $A(x)$, given by $A(x) = 2/(1 + e^{-x}) - 1$. The activation at the output layer was linear. It has been shown[25] that such a configuration satisfies the universal approximation theorem described previously.

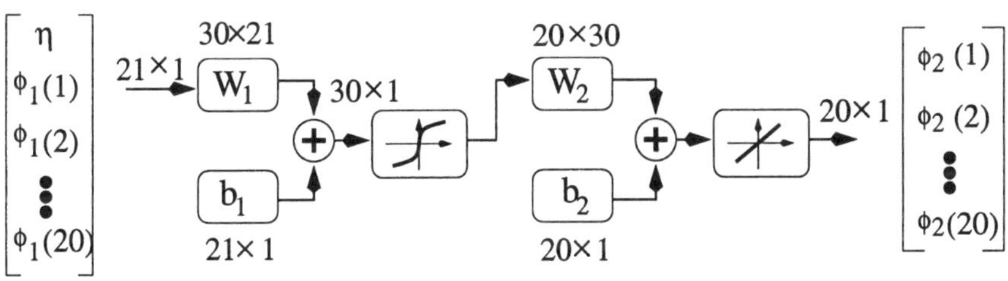

Fig. 4. Configuration of the ANN. W_1 and W_2 are weight matrices, and b_1 and b_2 are bias vectors. The activation functions (bipolar sigmoid function for the hidden layer, and linear function for the output layer) are illustrated in each box after the add sign $\oplus$. The dimensions of matrices and vectors, as well as the contents of the input/out vectors, are also indicated. *(Reused figure with permission from [17] © 2000 IEEE).*

The 21-element input vector, $\mathbf{V}_{in}$, to the ANN contained both the shape parameter, η, and ϕ_1, i.e., $\mathbf{V}_{in} = (\eta, \phi_1)$, while the 20-element output target vector, $\mathbf{V}_{out}$, consists of ϕ_2 only, i.e., $\mathbf{V}_{out} = \phi_2$. Prior to the training process, we normalized both $\mathbf{V}_{in}$ and $\mathbf{V}_{out}$ to the unit variance with respect to each element within these vectors (computed from 12,000 pairs of $\mathbf{V}_{in}$ and $\mathbf{V}_{out}$). This normalization results in an appropriate operating range which enables the ANN to be trained more efficiently.

7 Results

The backpropagation ANN was implemented on a 9000/802 HP workstation using the Neural Network Toolbox in the Matlab software package (Mathworks Inc., Natick, MA). Repeated experiments were performed to determine the sizes of hidden neurons and training samples. Our final ANN consists of 30 hidden neurons which provide a compromise between the mapping error and the computational cost. With this design the ratio between the number of training patterns (12,000) and the number of weights (1,230) was close to

10:1, as suggested by a commonly used rule of thumb[26]. Our experiments indicate that more training patterns provide a smaller mapping error and a better generalization; however, a larger dynamic memory (swap space in terms of the UNIX operating system) is required when the rapid batch training method is employed[27, 28]. During training the early stopping criterion was utilized which stopped training when the validation error (computed based on an independent 5,000-pattern validation set) started to increase. We explored various training algorithms, such as the resilient backpropagation[29] and four types of conjugate gradient algorithms (details are provided in [27, 28]). The results are shown in Table 1 with respect to the training algorithm, training time, number of epochs presented, relative training error, and relative test error. These relative mean-square errors are defined by

$$E = \frac{\sum_{i=1}^{M}(\phi_2(i) - \hat{\phi}_2(i))^2}{\sum_{i=1}^{M}\phi_2(i)^2} \times 100\% \tag{36}$$

where ϕ_2 and $\hat{\phi}_2$ are, respectively, the directly computed and ANN produced potentials for the spheroidal model, and $M = 12,000$ for the training case, and $M = 5,000$ for the test case.

Table 1. Results of ANN performance using ϕ_2 as the target vector

Training Method[27, 28]	Training Time	Epochs Presented	Relative Training Error	Relative Test Error
Resilient BP	573	4983	0.0039	0.0040
Scaled Conjugate	274	1295	0.0037	0.0039
Fletcher-Reeves	211	1091	0.0048	0.0051
Polak-Ribiere	255	1286	0.0039	0.0041
Powell-Beale	122	610	0.0043	0.0045

7.1 Alternative Design of Target Pattern Vector

In the previous design $\mathbf{V}_{in}$ and $\mathbf{V}_{out}$ are highly correlated because a spheroid with eccentricity $\eta \in [0.4, 0.6]$ is not greatly different from a sphere. As a result, we have $\mathbf{V}_{in}(i) \approx \mathbf{V}_{out}(i)$, for $i = 1, 2, \cdots, 20$, and the ANN primarily approximates the identical function. In order for the ANN to emphasize the difference between the two models, we re-defined the target training vector as $\mathbf{V'}_{out}(i) = \phi_2(i) - \phi_1(i)$, for $i = 1, 2, \cdots, 20$, i.e., $\phi_1(i)$ was used to predict $\phi_2(i)$. As a result, the identity component was removed from the mapping function, and the average amplitude $|\mathbf{V'}_{out}|$ became much smaller than $|\mathbf{V}_{out}|$. As in the previous case, we normalized both $\mathbf{V}_{in}$ and $\mathbf{V'}_{out}$ to the unit variance to improve the sensitivity of ANN to the model difference. It is clear that these procedures can be easily reversed to recover the original $\mathbf{V}_{out}$ from $\mathbf{V'}_{out}$.

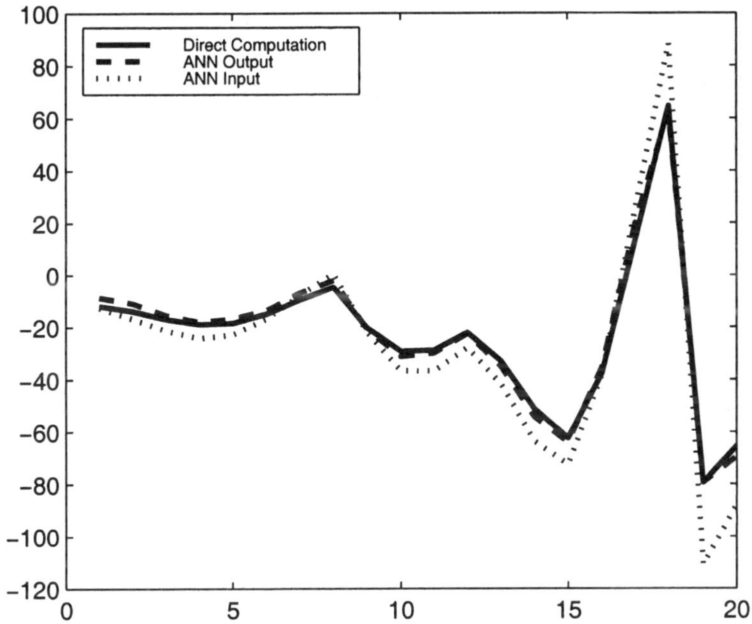

Fig. 5. A comparison among the ANN computed potentials (dash curve), the directly evaluated potentials (solid curve), and the input to the ANN (dot curve). The horizontal and vertical axes represent, respectively, channel number and potential value (in μv). In this comparison the relative mean-squared error is 0.0046. *(Reused figure with permission from [17] @ 2000 IEEE).*

The modified target vectors were utilized to train an ANN of the same configuration as in the previous case. The results are listed in Table 2. It can be observed that both the training and test relative errors have been improved significantly, and the training time has been shortened. Fig. 5 compares a particular forward solution (dashed curve) computed by the ANN to that by direct evaluation (solid curve). The input to the ANN is plotted by the dotted curve. This example was selected from the 5,000 independent test set and the Powell-Beale training algorithm[27, 28] was utilized. It can be seen that the solid and dashed curves are very close, indicating a close approximation by the ANN.

7.2 Computational Cost/Storage

The training for the ANN listed on Tables 1 and 2 requires several hours which are not overwhelming for an off-line, unattended computation. Once the ANN is trained, the on-line computation can be performed very rapidly as shown by the following measure of the number of floating point operations (flops). The total flops required consist of two major components: 1) those required

Table 2. Results of ANN performance using $\phi_2 - \phi_1$ as the target vector

Training Method[27, 28]	Training Time (min.)	Epochs Presented	Relative Training Error	Relative Test Error
Resilient BP	200	1739	0.0029	0.0031
Scaled Conjugate	70	330	0.0030	0.0032
Fletcher-Reeves	43	245	0.0031	0.0033
Polak-Ribiere	240	1189	0.0026	0.0028
Powell-Beale	95	453	0.0028	0.0030

for evaluating (8) which are estimated to be 45 flops per channel, or 900 flops for 20 channels; and 2) those required for evaluating the ANN in the form of multiplications between weight matrices and data vectors, where an $M \times N$ matrix multiplying an M-dimensional vector requires $(2M-1)N$ flops. In our case the ANN has a 21:30:20 configuration. The flops required to evaluate the ANN are approximately $21 \times 2 \times 30 + 30 \times 2 \times 20 = 2,460$ flops. Therefore, to compute 20-channels of forward solutions, the total computational cost is only $900 + 2460 = 3,360$ flops. If real-time processing is desired, these flops must be accomplished within 5 ms (assuming a sampling rate 200 Hz). This task represents little problem since a 400 MHz PC is capable of accomplishing 100,000 flops in 5 ms under a very conservative estimate of 20 clock cycles per flop.

The trained network requires storage for weights and biases. In our case we must store $21 \times 30 + 30 \times 20 = 1,230$ values for the weights and $30 + 20 = 50$ values for the biases (Fig. 4). Assuming each value is stored in the double precision floating-point format consisting of 8 bytes, the total storage required is only about 10.2 K-bytes.

8 Discussion

We have presented a soft-computing approach to forward solutions of the EEG based on non-spherical head models. We have also theoretically investigated the functional mapping between solutions to the Poisson PDE under fixed boundary surfaces. Two important results have been demonstrated: 1) the solution to the Poisson PDE is unique; and 2) the solution is continuous with respect to the variation of boundary surface. These theoretical results establish the validity of using a function approximation neural network to compute PDE solutions with any boundary surfaces that can be generalized from a canonical surface. When compared to the traditional finite element and boundary element methods, the neural network method is knowledge-based, robust, and efficient. Our experimental study on functional mapping between PDE solutions has indicated that soft-computing can significantly reduce computational costs, allow real-time implementation, and produce highly accurate numerical results.

Since there exists a class of other PDEs that possess similar uniqueness and continuity to the Poisson PDE, the theoretical results and computational methods presented in this paper can be generalized. We believe that the ANN soft-computing method provides a powerful alternative to the traditional FEM and BEM methods for a variety of practical applications.

9 Acknowledgment

This work was supported by National Institutes of Health grants No. NS-38494, EB-002309, and EB-002099, and by Computational Diagnostics, Inc..

APPENDIX: Mapping Points between the Spherical and Spheroidal Spaces

The mapping between electrode positions $\mathbf{r} = \{\hat{x}, \hat{y}, \hat{z}\}$ and $\mathbf{r}_\eta = \{\tilde{x}, \tilde{y}, \tilde{z}\}$ has been illustrated in Fig. 3. The line, denoted by l, passes through points c and $\mathbf{e}$. Hence, its line equation is given by

$$\frac{x}{\sin\psi_0 \cos\theta_0} = \frac{y}{\sin\psi_0 \sin\theta_0} = \frac{z}{\cos\psi_0} \tag{37}$$

where ψ_0 and θ_0 are, respectively, the azimuth and elevation angles that represent $\mathbf{r}$ on the unit sphere. By the definition of mapping, line l must also pass $\mathbf{e}_\eta$, we then have

$$\frac{\sin\psi_1 \cos\theta_1}{\sin\psi_0 \cos\theta_0} = \frac{\sin\psi_1 \sin\theta_1}{\sin\psi_0 \sin\theta_0} = \frac{a\cos\psi_1}{\cos\psi_0} \tag{38}$$

where ψ_1 and θ_1 are, respectively, the azimuth and elevation angles that represent $\mathbf{r}_\eta$. Note that, in (38), we have assumed that the short axis of the spheroid is equal to one (see Section 6.3).

Solving (38), we have

$$\theta_1 = \theta_0$$

and

$$\tan\psi_1 = a\tan\psi_0. \tag{39}$$

By the nature of the problem, line l generally passes through the surface of the spheroid twice. Thus, there are two candidates for $\mathbf{r}_\eta$. The desired candidate is the one with a smaller $|\mathbf{r} - \mathbf{r}_\eta|$. Using the above results, we can now explicitly express $\mathbf{r}_\eta$ by

$$\begin{aligned} \tilde{x} &= \sin\psi_1 \cos\theta_0 \\ \tilde{y} &= \sin\psi_1 \sin\theta_0 \\ \tilde{z} &= a\cos\psi_1 \end{aligned} \tag{40}$$

with

$$\psi_1 = \begin{cases} \psi_0, & \text{for } \psi_0 = \pm\frac{\pi}{2}; \\ \tan^{-1}(a \tan \psi_0) + k\pi, & \text{otherwise} \end{cases} \tag{41}$$

where $k \in \{0, 1\}$. The case for mapping the two dipole positions $\mathbf{r}'$ and $\mathbf{r}'_\eta$ (see Fig. 3) can be similarly derived. The results are the same as shown in (40) and (41) except that $\tilde{x}$, $\tilde{y}$, and $\tilde{z}$ are all multiplied by the modulus $|\mathbf{r}'|$.

References

1. Fischbach GD (1992), "Mind and Brain,." *Scientific American,* **267**, 48-57.
2. Swartz BE, Goldensohn ES (1998), "Timeline of the history of EEG and associated fields," *Electroencephalogr Clin. Neurophysiol.*, **106**, 173-176.
3. Malmivuo J and Plonsey R (1995), *Bioelectromagnetism*, Oxford University Press, Oxford, UK.
4. Hamalainen M, Hari R, Ilmoniemi RJ, et al (1993), "Magnetoencephalography - theory, instrumentation, and applications to noninvasive studies of the working human brain," *Reviews of Modern Physics,* **65**, 413-497.
5. Gulrajani RM (1998), *Bioelectricity and Biomagnetism*, John Wiley and Sons, New York, NY.
6. Jackson JD (1999), *Classical Electrodynamics*, John Wiley & Sons, 4th ed., New York, NY.
7. Spiegel M. (1968), *Schaum's Outline of Vector Analysis,* McGraw-Hill, New York, NY.
8. Sidman RD, Giambalvo V, Allison T, and Bergey P (1978), "A method for localization of sources of human cerebral potentials evoked by sensory stimuli," *Sensory Processes,* **2**, 116-129.
9. Salu Y, Cohen LG, Rose D, Sato S, Kufta C, and Hallett M (1990), "An improved method for localizing electric brain dipoles," *IEEE Trans. Biomed. Engr.,* **37**, 699-705.
10. Sun M (1997), "An efficient algorithm for computing multishell spherical head models for EEG source localization," *IEEE Trans. Biomed. Eng.*, **44**, 1243-1252.
11. Cuffin BN, Cohen D (1979), "Comparison of the magnetoencephalogram and electroencephalogram," *Electroencephalogr Clin Neurophysiol,* **47**, 132-146.
12. Berg P. and Scherg M (1994), "A fast method for forward computation of multiple-shell spherical head models," *Electroencephalogr Clin Neurophysiol,* **90**, 58-64.
13. Sun M (1997), "Computing the forward EEG solution of the multishell spherical head model for localizing electrical activity in the brain," *In Proc. IEEE EMBC'97,* Chicago, 1172-1175.
14. Yeh GCK and Martinek J (1956), "The potential of a general dipole in a homogeneous conducting spheroid," *Ann. N. Y. Acad. Science* **65**, 1003-1006.
15. Munck JC de (1988), "The potential distribution in a layered anisotropic spheroidal volume conductor," *J. Appl. Phys.*, **64**, 464-470.
16. Fender DH (1991), "Models of the human brain and the surrounding media: their influence on the reliability of source localization," *J. Clinical Neurophysiology,* **8**, 381-390.

17. Sun M and Sclabassi RJ (2000), "The Forward EEG Solutions Can be Computed Using Artificial Neural Networks," *IEEE Transactions on Biomedical Engineering,* **47**, 1044-1050.
18. Sclabassi RJ, Sonmez M, and Sun M (2001), "EEG source localization: a neural network approach," *Neurological Research,* **23**, 457-464.
19. Sun M, Yan X, and Sclabassi RJ (2003), "Solving Partial Differential Equations in Real-Time Using Artifical Neural Network Signal Processing as an Alternative to Finite Element Analysis," *In Proc. IEEE ICNNSP'03,* Nanjing, China, 381-384.
20. Haykin S (1994), *Neural Networks: A Comprehensive Foundation,* Maxwell Macmillan Canada, Toronto.
21. Zwillinger D (1996), *Standard mathematical tables and formulae* , CRC Press, Boca Raton, FL.
22. N. S. Trudinger (2001), *Elliptic Partial Differential Equations Of Second Order,* Springer-Verlag, Berlin.
23. Mathematical Tables Project (1945), *Tables of associated Legendre functions.* Conducted under the sponsorship of the National Bureau of Standard, Columbia University Press, New York.
24. Bocker KBE, van Avermaete JAG, and van den Berg-Lenssen MMC (1994), "The international 10-20 system revised: Cartesian and spherical co-ordinates," *Brain Topography,* **6**, 231-235.
25. Fausett L (1994), "Fundamentals of neural networks," Prentice-Hall, Inc., Englewood Cliffs, New Jersey.
26. Swingler K (1996), *Applying neural networks,* Academic Press, San Diego, CA.
27. User's Guide Manual (1998): *Neural Network Toolbox,* Version 3, Mathworks, Inc., Matick, MA.
28. Hagan, MT, Demuth HB, Beale MH (1996), *Neural network design,* PWS Publishing, Boston, MA.
29. Riedmiller M, Braun H (1993), "A direct adaptive method for faster backpropagation learning: the RPROP algorithm," *Proc. IEEE Int. Conf. on Neural Networks.*

Providing Common Time and Space in Distributed AV-Sensor Networks by Self-Calibration

R. Lienhart[1], I. Kozintsev[1], D. Budnikov[2], I. Chikalov[2], and V. C. Raykar[3]

[1] Intel Research, Intel Corporation, 2200 Mission College Blvd, Santa Clara, CA 95052, USA Rainer.Lienhart@intel.com
[2] Intel Research, Intel Corporation, Turgeneva st., 30, Nizhny Novgorod, Russia
[3] Perceptual Interfaces and Realities Lab., University of Maryland, College Park, USA

Abstract. Array audio-visual signal processing algorithms require time-synchronized capture of AV-data on distributed platforms. In addition, the geometry of the array of cameras, microphones, speakers and displays is often required. In this chapter we present a novel setup involving network of wireless computing platforms with sensors and actuators onboard, and algorithms that can provide both synchronized I/O and self-localization of the I/O devices in 3D space. The proposed algorithms synchronize input and output for a network of distributed multi-channel audio sensors and actuators connected to general purpose computing platforms (GPCs) such as laptops, PDAs and tablets. IEEE 802.11 wireless network is used to deliver the global clock to distributed GPCs, while the interrupt timestamping mechanism is employed to distribute the clock between I/O devices. Experimental results demonstrate a precision in A/D D/A synchronization precision better than 50 μs (a couple of samples at 48 kHz). We also present a novel algorithm to automatically determine the relative 3D positions of the sensors and actuators connected to GPCs. A closed form approximate solution is derived using the technique of metric multidimensional scaling, which is further refined by minimizing a non-linear error function. Our formulation and solution account for the errors in localization, due to lack of temporal synchronization among different platforms. The performance limit for the sensor positions is analyzed with respect to the number of sensors and actuators as well as their geometry. Simulation results are reported together with a discussion of the practical issues in a real-time system.

Keywords: distributed sensor networks, self-localizing sensor networks, multichannel signal processing

1 Introduction

Arrays of audio/video sensors and actuators such as microphones, cameras, loudspeakers and displays along with array processing algorithms offer a rich set of new features for emerging applications. Until now, array processing required expensive dedicated multi-channel I/O cards and high-throughput computing systems to process multiple channels on a single machine. Recent advances in mobile computing and communication technologies, however, suggest a novel and very attractive platform for implementing these algorithms. Students in classrooms and co-workers at meetings are nowadays accompanied by several mobile computing and communication devices with audio and video I/O capabilities onboard such as laptops, PDA's, and tablets. In addition, high-speed wireless network connections, like IEEE 802.11a/b/g, are available to network those devices. Such ad-hoc sensor/actuator networks can enable emerging applications that include multi-stream audio and video, smart audio/video conference rooms, meeting recordings, automatic lecture summarization, hands-free voice communication, speech enhancement and object localization. No dedicated infrastructure in terms of the sensors, actuators, multi-channel interface cards and computing power is required. Multiple GPCs along with their sensors and actuators co-operate on providing transparent synchronized I/O. However, there are several important technical and theoretical problems to be addressed before the idea of using those devices for array DSP algorithms can materialize in real-life applications.

Fig. 1. Distributed computing platform consisting of N general-purpose computers along with their onboard audio sensors, actuators and wireless communication capabilities.

Fig. 1 shows a schematic representation of our proposed *distributed computing platform* consisting of N GPC platforms(e.g., laptops). Each GPC is equipped with audio sensors (microphones), actuators (loudspeakers), and wireless communication capabilities. Given this setup, one of the most important problems is to provide a common reference time to a network of distributed computers and their I/O channels. A second important problem is to provide a common 3D coordinate system for the locations of the sensors and actuators. Solutions to both problems will be presented.

2 Providing Common Time

To illustrate the importance of time synchronization, we implemented a Blind Source Separation (BSS) algorithm published in [2]. In the simplest setting, two sound sources are separated using the input of two microphones, each connected to a different laptop. However, without synchronization of A/Ds the BSS algorithm failed to perform separation. Fig. 2 demonstrates how a difference of only a few Hz in audio sampling frequency between two channels (laptops) impacts source separation. On the x-axis the sampling difference in Hz between two audio channels at about 16 kHz is shown against the achieved signal separation gain by BSS on the y-axis. As can be seen in Fig. 2, a difference of only 2 Hz at 16 kHz reduces the signal separation gain from 8.5 dB to about 2 dB only. In real life the difference in sampling frequency can be even higher as we illustrate in Table 1. BSS is not the only algorithm that is extremely sensitive to sampling synchronization. Other applications that require similar precision of time synchronization between channels are acoustic beamforming and 3D audio rendering.

Table 1. Audio sampling rates of several laptops

Laptop	Dell Inspiron 7000	IBM ThinkPad T20	IBM ThinkPad 600E	IBM ThinkPad T23
Sampling rate, Hz	16001.7	16003.6	16001.8	16009.5

2.1 Related Work

The problem of time synchronization in distributed computing systems has been discussed extensively in the literature in the context of maintaining clock synchrony throughout large geographic areas. Each process exchanges messages with its peers to determine a common clock. Seminal works have been reported in [6] and [9]. However, the results provided there can not be applied

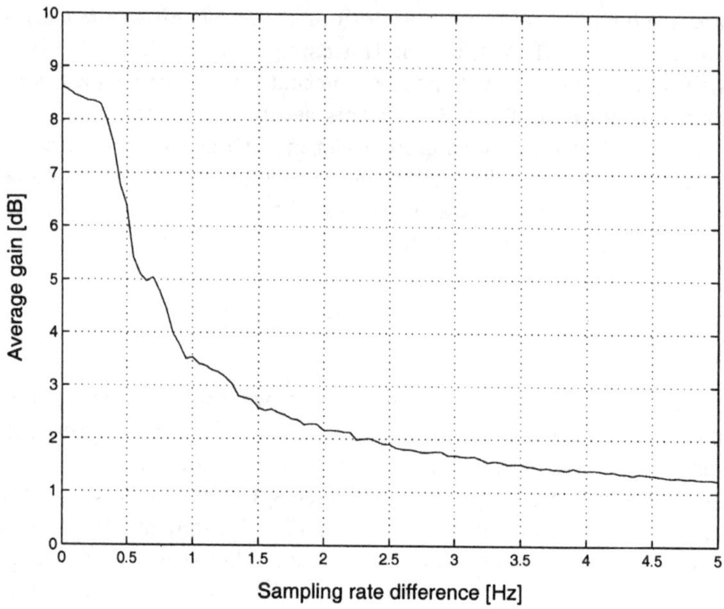

Fig. 2. Sensitivity of acoustic source separation performance to small sample rate differences. Channel 1 is assumed to sample at 16 kHz, while channel 2 is assumed to sample at $16000 + x$ Hz. Signal separation gain is calculated for the Blind Source Separation algorithm in [2].

directly to our problem, since the precision of time synchronization is too low. NTP, the Network Time Protocol, currently used worldwide for clock synchronization in the best case achieves synchronization in the range of milliseconds - 2 to 3 orders of magnitudes higher than the microsecond resolution needed for our application scenarios. The Global Positioning system (GPS) provides a much higher clock resolution. Its reported time is steered to stay always within one microsecond of UTC (Coordinated Universal Time). In practice, it has been within 50 nanoseconds. With the Standard Positioning Service (SPS) a GPS receiver can obtain a time transfer accuracy to UTC within 340 nanoseconds (95% interval). GPS, however, only works reliably outdoors and thus does not completely fit our application scenario. There is also some recent work on synchronization in wireless sensor networks. In [10, 1], the reference-broadcast synchronization method is introduced. In this scheme, nodes send reference beacons to their neighbors based on a physical broadcast medium. All nodes record the local time at which they receive the broadcasts (e.g., by using the RDTSC instruction of the Pentium® processor family; the Read-Time Stamp Counter counts clocktics since the processor was started). Based on the exchange of this information, nodes can translate each other's clock. Although promising, the worst case performance of 150μs reported in [10] is too high for our application scenario. Our system is similar in spirit but we rely on additional processing to reduce errors in estimation of synchronization parameters.

In general, all clock synchronization algorithms studied in the literature only address the problem of providing a common clock on distributed computing platforms. They do not address how the I/O can be synchronized with the common clock (we proposed one solution in [7]). In other words, even under the assumption of a perfect clock on each platform, there is still a mechanism required to link the common clock to the data in the I/O channels. On a GPC this is a challenge in itself and we address this problem in this chapter.

2.2 Problem Formulation

We tackle the problem of distributed I/O synchronization in two steps: (1) the local CPU clocks of the GPCs are synchronized against a global clock (inter-platform), and (2) I/O is synchronized against the local clocks and thus also against the global clock (intra-platform). In the experimental results, one of the CPU clocks will arbitrarily be chosen as the global clock.

Each GPC has a local CPU clock (e.g., RDTSC). Let $t_i(t)$ denote the value of this clock on the i-th GPC at some global time t. Assuming a linear model between the global clock and the local platform clock, we get

$$t_i(t) = a_i(t)t + b_i(t), \tag{1}$$

where $a_i(t)$ and $b_i(t)$ are timing model parameters for the i-th GPC. The dependency of the model parameters on global time t approximates instabilities in the clock frequency due to temperature variations and other factors. In practice, these instabilities are in the order of 10^{-5}. In the rest of this section we will omit explicit time dependency to simplify our notations. Similarly, the sampling times of audio A/Ds and D/As on GPC's are approximated as:

$$\tau_i(t_i) = \alpha_i(t_i)t_i + \beta_i(t_i). \tag{2}$$

In this model τ_i is simply the number of samples produced by A/D (or consumed by D/A) converter since the start of the audio I/O. Note that two different timing models are required since the audio I/O devices on a typical PC platform have their own internal clock that is not synchronized to other platform clocks such as the RDTSC.

Given the two timing models above the problem that we address in this section can be formulated as finding $t(\tau_i)$ - the global time stamp of audio sample τ_i. We separate it into two subproblems: finding $\hat{\alpha}_i$ and $\hat{\beta}_i$ such that $t_i(\tau_i) = \hat{\alpha}_i\tau_i + \hat{\beta}_i$ (convert sample number to local time stamp with $\hat{\alpha} = \alpha^{-1}$ and $\hat{\beta} = -\beta/\alpha$) and finding $\hat{a}$ and $\hat{b}$ such that $t(t_i) = \hat{a}t_i + \hat{b}$ (convert value of local clock to global time with $\hat{a} = a^{-1}$ and $\hat{b} = -b/a$).

2.3 Timing Relationships on GPC Platform

In order to understand the inter and intra platform synchronization methods proposed here we briefly describe the operations and timing relationships on a typical GPC. Fig. 3 shows a processing diagram of networking and audio

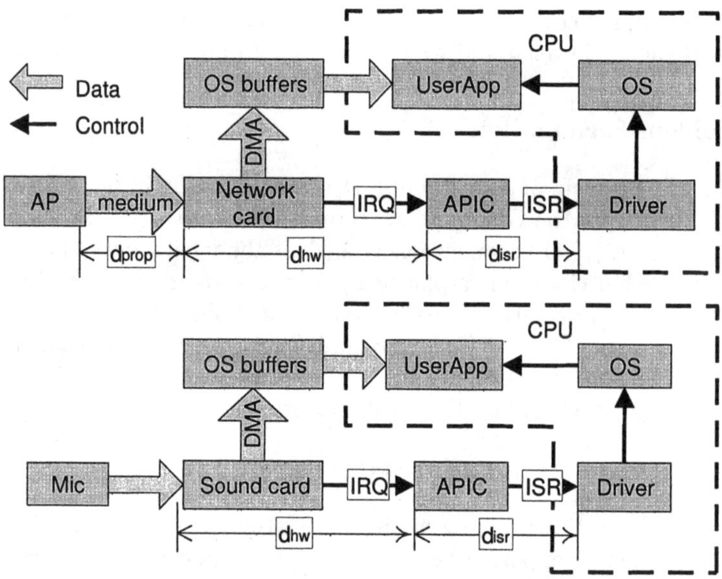

Fig. 3. Network (top part) and audio (bottom part) data and control flows on a typical GPC platform.

I/O. Both I/O operations have a very similar structure that can be described by the following sequence of actions (only input path is described):

1. Incoming data is received and processed by a hardware device, and eventually is put into a Direct Memory Access (DMA) buffer. This is modeled in Fig. 3 by the delay d_{hw}, which is approximately constant for similar hardware.
2. The DMA controller transfers the data to a memory block allocated by the system and signals this event to the CPU by an Interrupt ReQuest (IRQ). This stage introduces variable delay due to memory bus arbitration between different agents (i.e., CPU, graphics adapter, other DMA's).
3. The interrupt controller (APIC) queues the interrupt and schedules a time slot for handling. Because APIC is handling requests from multiple I/O devices this stage introduces variable delay with standard deviation of around 6 ms and the maximum deviation of 30 ms. Both previous stages are modeled by d_{isr} in Fig. 3.

4. The Interrupt Service Routine (ISR) of the device driver is called, and the driver sends notification to the Operating System (OS).
5. The OS delivers a notification and data to the user application(s). This stage has to be executed in a multitasking software environment and this leads to significant variable delays that depend on CPU utilization and many other factors.

In summary, data traverses multiple hardware and software stages in order to travel from an I/O device to the CPU and back. The delay introduced by the various stages is highly variable making the problem of providing a global clock to the GPCs and distributing it to I/O devices very challenging. It is advantageous to perform synchronization as close to hardware as possible, therefore our solution is implemented at the driver level (during ISR) thus avoiding additional errors due to OS processing.

2.4 Inter-platform synchronization

For the synchronization of CPU clocks over a wireless network we propose to use a series of arrival times of multicast packets sent by the wireless access point (AP). In our current approach we implement a pairwise time synchronization with one node chosen as the master (say $t(t_0) = t_0$). All other nodes (clients) are required to synchronize their clocks to the master . A similar approach was also suggested in [10, 1]. Our solution, however, extends it by introducing additional constraints on the timing model. In order to provide a global clock to distributed platforms that is potentially useful to other applications (e.g., joint stream processing and distributed computations), we impose the clock monotonicity condition to make sure that the global clock is monotonically increasing during model parameter adaptation. In addition we smooth the clock model (a_i and b_i in Eq. (1)) variation by limiting the magnitude of its updates.

The algorithm consists of the following steps:

1. AP sends next beacon packet.
2. Master node records its local time of packet arrival and distributes it to all other nodes.
3. Client nodes record both their local times of arrival of beacon packets from AP, and the corresponding times received from the master.
4. Clients update local timing models based on the set of local timestamps and corresponding master timestamps.

Let us assume that in Fig. 3 the packet j arrives to multiple platforms approximately at the same global time corresponding to local clocks t_i^j ($d_{prop} \approx 0$). The set of observations available on the platforms consist of pairs of timestamps $(\tilde{t}_0^j, \tilde{t}_i^j)$. From Fig. 3 we have $\tilde{t}^j = t^j + d_{hw} + d_{isr}$ (we omitted dependency on i) that we further model as $\tilde{t}^j = t^j + d + n$. In this approximation d models all constant delay component and n represents the stochastic component. Given the set of observations $(\tilde{t}_0^j, \tilde{t}_i^j)$ we are required to estimate the timing model parameters $\hat{a}_i$ and $\hat{b}_i$ for all slave platforms. In our experiments a window of 3 minutes is used to estimate current values of $\hat{a}_i$ and $\hat{b}_i$ using the least trimmed squares (LTS) regression [14]. LTS is equivalent to performing least squares fit, trimming the observations that correspond to the largest residuals (defined as the distance of the observed value to the linear fit), and then computing a least squares regression model for the remaining observations. Fig. 4 shows comparison of quantiles of residuals with quantiles of normal distribution and Fig. 5 plots the histogram of residuals. The distribution appears to be close to Gaussian except for the presence of a few outliers (see Fig. 4) that do not fit into a normal distribution. The trimming step is specifically targeted to remove those outliers.

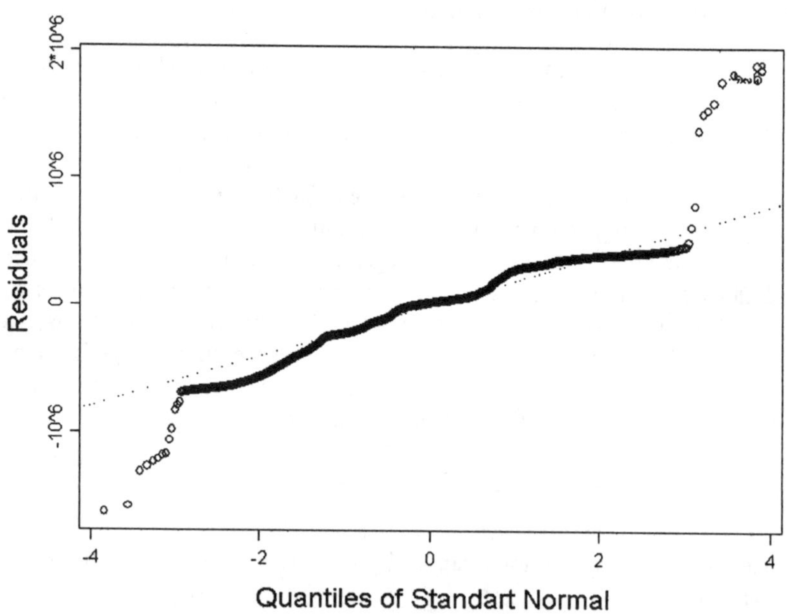

Fig. 4. Comparison of quantiles of residuals with quantiles of the normal distribution. Points away from the straight line are treated as outliers and removed during regression.

2.5 Intra-platform synchronization

In order to synchronize the audio clock to the CPU clock we use a similar approach as the one presented in the previous section. The ISR of the audio

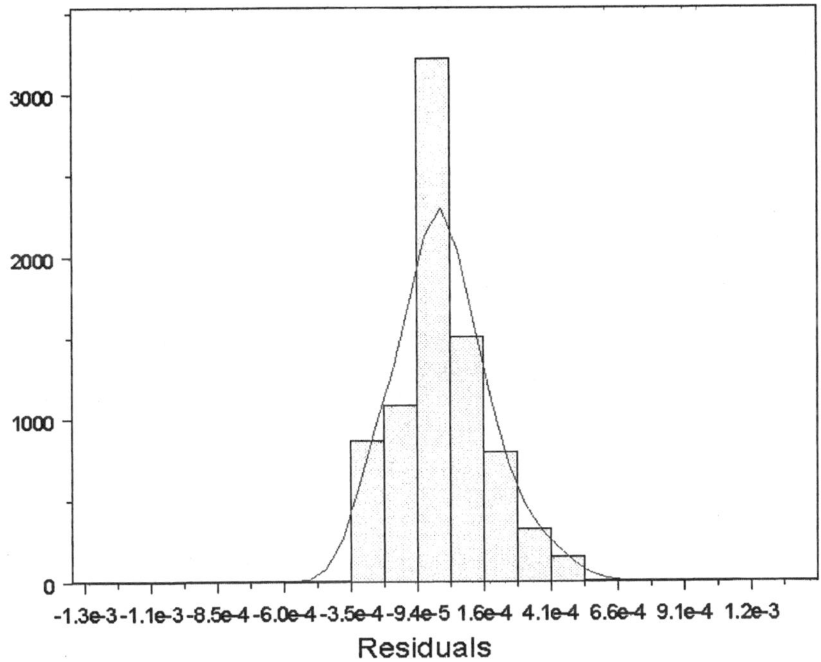

Fig. 5. Histogram of residuals and the normal probability density function.

driver is modified to timestamp the samples in the OS buffer using the CPU clock to form a set of observation pairs $(\tilde{t}_i^j, \tau_i^j)$, where j now represents the index of an audio data packet.

Following our model in Fig. 3 we have $\tilde{t}^j = t^j + d_{hw} + d_{isr}$ (we omitted dependency on i) that we further represent as $\tilde{t}^j = t^j + d + n$. Except for the fact that the τ^j is available without any noise (it is simply the number of samples processed!) we are back to the problem of determining the linear fit parameters for pairs of observations that we solved in the previous section using the LTS method.

In summary, by using LTS procedure twice both local and global synchronization problems are solved and the audio samples can be precisely synchronized on the distributed GPCs.

2.6 Experimental results

The distributed test system was implemented with several off-the-shelf Intel® Centrino laptops using the following software components (see also Fig. 6): (a) A *modified WLAN card driver* timestamps each interrupt, parses incoming packets in order to find all master beacon frames, and stores their timestamp values in a cyclic shared memory buffer. The timestamp values as well as the corresponding message IDs are further accessible through the standard driver I/O interface. (b) A *modified AC97 driver* timestamps ISRs and calculates the number of samples transmitted since the beginning of the audio capture/playblack. The value pair is placed into a cyclic shared memory buffer. (c) The *synchronization agents* are responsible for synchronizing the distributed system. We have three types of agents: the multicast server (MCS), the master synchronization agent (SAM) and the slave synchronization agent (SAS). The MCS periodically broadcasts beacon packets (short packets with unique ID as the payload). The SAM and SASs use the modified WLAN driver to detect the beacons. The SAM periodically broadcasts its recorded timestamps of beacon arrivals to the SAS devices. Based on SASs' recorded timestamps and the corresponding SAM timestamps, each SAS calculates the clock parameter to convert between the platform clock and the global clock. The clock parameters are placed in shared memory for use by other applications. (d) The *Synchronization API* allows user applications to retrieve the local clock value, access the clock parameters, and convert between the platform and global clock. (e) The *audio API* allows user applications to retrieve pairs of local timestamps and sample numbers, as well as to convert global timestamp values to sample numbers and vice versa. It also provides transparent synchronized capture and playback.

Based on these components a distributed audio rendering system was implemented with three laptops (see Fig. 6). The first laptop was used as the MCS. Modified AC97 and WLAN drivers were installed on other two laptops. SAM was started on the second laptop, while SAS was started on the third laptop. The distributed system was instructed through the audio API to synchronously playback a Maximum Length Sequence (MLS) signal on the two synchronized laptops. The line-out signal of both laptops were recorded by a multichannel soundcard. The measured inter-GPC offset was at most 2 samples at 48 kHz (less than 42 μs).

3 Providing Common Space

A common space (coordinate system) can be provided by means of actively estimating the three dimensional positions of the sensors and actuators. Many multi-microphone array processing algorithms (like sound source localization or conventional beamforming) need to know the positions of the microphones very precisely. Current systems either place the microphones in known locations or manually calibrate them. There are some approaches which do calibration using speakers in known locations [15]. We offer here a more general approach where no assumptions about the positions of the speakers are made. Our solution explicitly accounts for the errors in localization due to lack of temporal synchronization among different platforms.

We again refer to Fig. 1 showing a schematic representation of a distributed computing platform consisting of N GPCs. For the purpose of performing

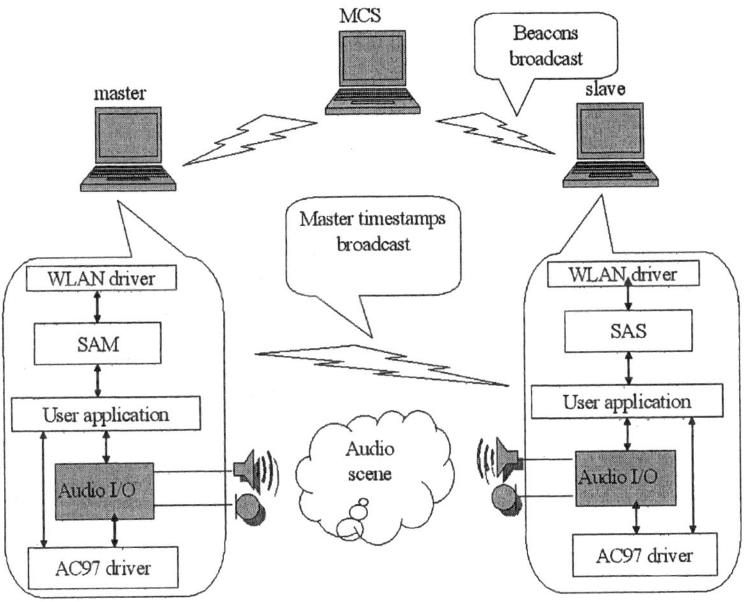

Fig. 6. Distributed audio rendering/capturing system setup

space localization one of them is configured to be the master. The master controls the distributed computing platform and performs the location estimation. As already described each GPC is assumed to be equipped with audio sensors (microphones), actuators (loudspeakers), and wireless communication capabilities.

3.1 Related Work

The problem of self-localization for a network of nodes generally involves two steps: ranging and multilateration. The ranging technology can be either based on the Time Of Flight (TOF) or the Received Signal Strength (RSS) of acoustic, ultrasound or radio frequency (RF) signals. The Global Positioning System (GPS) system and long range wireless sensor networks use RF technology for range estimation. Localization using GPS is not suitable for our applications since GPS systems do not work indoors and are very expensive. Also RSS based on RF is very unpredictable [16] and the RF TOF is quite small to be used indoors. [16] discusses systems based on ultrasound TOF using specialized hardware (like motes) as the nodes. However, our goal is to use the already available sensors and actuators on GPCs to estimate their positions. Our ranging technology is based on acoustic TOF as in [15, 11, 4]. Once we have the range estimates the Maximum Likelihood (ML) estimate can be used to get the positions. To find the solution one can assume that the locations of a few sources are known as in [15, 16] or make no such assumptions as in [11, 19].

3.2 Problem Formulation

Given a set of M acoustic sensors (microphones) and S acoustic actuators (speakers) in unknown locations, our goal is to estimate their three dimensional coordinates. Each of the acoustic actuators is excited using a known calibration signal such as maximum length sequences or chirp signals, and the Time of Flight (TOF) is estimated for each of the acoustic sensors. The TOF for a given pair of microphone and speaker is defined as the time taken by the acoustic signal to travel form the speaker to the microphone.

Let $\mathbf{m_i}$ for $i \in [1, M]$ and $\mathbf{s_j}$ for $j \in [1, S]$ be the three dimensional vectors representing the spatial coordinates of the ith microphone and j^{th} speaker, respectively. We excite one of the S speakers at a time and measure the TOF at each of the M microphones. Let TOF_{ij}^{actual} be the actual TOF for the ith microphone due to the jth source. Based on geometry the actual TOF can be written as (assuming a direct path),

$$TOF_{ij}^{actual} = \frac{\| \mathbf{m_i} - \mathbf{s_j} \|}{c}, \tag{3}$$

where c is the speed of sound in the acoustical medium [4] and $\|\|$ is the Euclidean norm. The TOF which we estimate based on the signal captured conforms to this model only when all the sensors start capturing at the same instant and we know when the calibration signal was sent from the speaker.

However in a typical distributed setup as shown in Fig. 1, the master starts the audio capture and playback on each of the GPCs one by one. As a result the capture starts at different instants on each GPC and also the time at which the calibration signal was emitted from each loudspeaker is not known. So the TOF which we measure from the signal captured includes both the speaker emission start time and the microphone capture start time (see Fig. 7 where $\hat{TOF}_{ij}$ is what we measure and TOF_{ij} is what we require). The speaker emission start time is defined as the time at which the sound is actually emitted from the speaker. This includes the time when the play back command was issued (with reference to some time origin), the network delay involved in starting the playback on a different machine (if the speaker is on a different GPC), the delay in setting up the audio buffers and also the time required for the speaker diaphragm to start vibrating. The microphone capture start time is defined as the time instant at which capture is started. This includes the time when the capture command was issued, the network

[4]The speed of sound in a given acoustical medium is assumed to be constant. In air it is given by $c = (331 + 0.6T)m/s$, where T is the temperature of the medium in Celsius degrees.

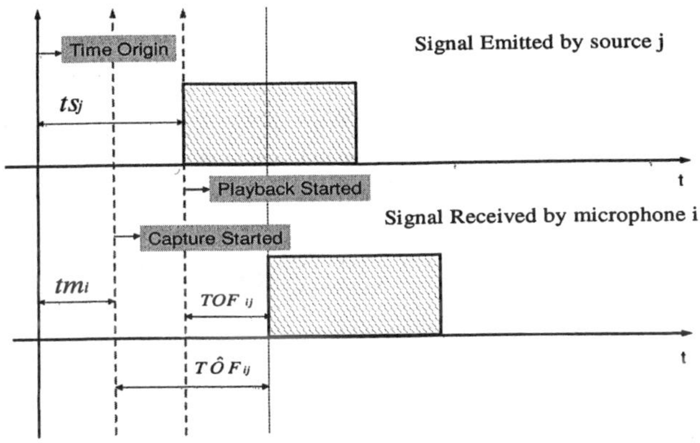

Fig. 7. Schematic indicating the errors due to unknown speaker emission and microphone capture start time.

delay involved in starting the capture on a different machine and the delay in transferring the captured sample from the sound card to the buffers.

Let ts_j be the emission start time for the jth source and tm_i be the capture start time for the ith microphone (see Fig. 7). Incorporating these two the actual TOF now becomes,

$$T\hat{O}F_{ij}^{actual} = \frac{\| \mathbf{m_i} - \mathbf{s_j} \|}{c} + ts_j - tm_i. \tag{4}$$

The origin can be arbitrary since $T\hat{O}F_{ij}^{actual}$ depends on the difference of ts_j and tm_i. We start the audio capture on each GPC one by one. We define the microphone on which the audio capture was started first as our first microphone. In practice, we set $tm_1 = 0$, i.e., the time at which the first microphone started capturing is our origin. We define all other times with respect to this origin. We can jointly estimate the unknown source emission and capture start times along with microphone and source coordinates.

In this chapter we propose to use the Time Difference Of Arrival (TDOA) instead of the TOF. The TDOA for a given pair of microphones and a speaker is defined as the time difference between the signal received by the two microphones [5]. Let $TDOA_{ikj}^{estimated}$ be the estimated TDOA between the ith and the kth microphones when the jth source is excited. Let $TDOA_{ikj}^{actual}$ be the actual TDOA. It is given by

$$TDOA_{ikj}^{actual} = \frac{\| \mathbf{m_i} - \mathbf{s_j} \| - \| \mathbf{m_k} - \mathbf{s_j} \|}{c}. \tag{5}$$

Including the source emission and capture start times, it becomes

[5] Given M microphones and S speakers we can have $MS(M-1)/2$ TDOA measurements as opposed to MS TOF measurements. Of these $MS(M-1)/2$ TDOA measurements only $(M-1)S$ are linearly independent.

$$\hat{TDOA}_{ikj}^{actual} = \frac{\| \mathbf{m_i} - \mathbf{s_j} \| - \| \mathbf{m_k} - \mathbf{s_j} \|}{c} + tm_k - tm_i. \quad (6)$$

In the case of TDOA the source emission time is the same for both microphones and thus gets canceled out. Therefore, by using TDOA measurements instead of TOF we can reduce the number of parameters to be estimated.

3.3 Maximum Likelihood (ML) Estimate

Assuming a Gaussian noise model for the TDOA observations, we can derive the ML estimate as follows. Let Θ be a vector of length $P \times 1$, representing all the unknown non-random parameters to be estimated (microphone and speaker coordinates and microphone capture start times). Let Γ be a vector of length $N \times 1$, representing noisy TDOA measurements. Let $T(\Theta)$ be a vector of length $N \times 1$, representing the actual values of the observations. Then our model for the observations is $\Gamma = T(\Theta) + \eta$ where η is the zero-mean additive white Gaussian noise vector of length $N \times 1$ where each element has the variance σ_j^2. Also let us define Σ to be the $N \times N$ covariance matrix of the noise vector N. The likelihood function of Γ in vector form can be written as:

$$p(\Gamma/\Theta) = (2\pi)^{-\frac{N}{2}} \mid \Sigma \mid^{-\frac{1}{2}} \exp -\frac{1}{2}(\Gamma - T)^T \Sigma^{-1}(\Gamma - T). \quad (7)$$

The ML estimate of Θ is the one which maximizes the log likelihood ratio and is given by

$$\hat{\Theta}_{ML} = \arg_{\Theta} \max F(\Theta, \Gamma),$$
$$F(\Theta, \Gamma) = -\frac{1}{2}[\Gamma - T(\Theta)]^T \Sigma^{-1}[\Gamma - T(\Theta)]. \quad (8)$$

Assuming that each of the TDOAs are independently corrupted by zero-mean additive white Gaussian noise [6] of variance σ_{ikj}^2 the ML estimate turns out to be a nonlinear least squares problem (in this case Σ is a diagonal matrix), i.e.,

$$\hat{\Theta}_{ML} = \arg_{\Theta} \min[\widetilde{F}_{ML}(\Theta, \Gamma)],$$
$$\widetilde{F}_{ML}(\Theta, \Gamma) =$$
$$\sum_{j=1}^{S}\sum_{i=1}^{M}\sum_{k=i+1}^{M} \frac{(TDOA_{ikj}^{estimated} - \hat{TDOA}_{ikj}^{actual})^2}{\sigma_{ikj}^2}. \quad (9)$$

---

[6]We estimate the TDOA or TOF using Generalized Cross Correlation (GCC)[5]. The estimated TDOA or TOF is corrupted due to ambient noise and room reverberation. For high SNR the delays estimated by the GCC can be shown to be normally distributed with zero mean [5].

Since the solution depends only on pairwise distances, any translation, rotation and reflection of the global minimum found will also be a global minimum. In order to make the solution invariant to rotation and translation we select three arbitrary nodes to lie in a plane such that the first is at $(0,0,0)$, the second at $(x_1,0,0)$, and the third at $(x_2,y_2,0)$. In two dimensions we select two nodes to lie in a line, the first at $(0,0)$ and the second at $(x_1,0)$. To eliminate the ambiguity due to reflection along Z-axis (3D) or Y-axis (2D) we specify one more node to lie in the positive Z-axis (in 3D) or positive Y-axis (in 2D). Also the reflections along X-axis and Y-axis (for 3D) can be eliminated by assuming the nodes which we fix to lie on the positive side of the respective axes, i.e., $x_1 > 0$ and $y_2 > 0$. Similar to fixing a reference coordinate system in space we introduce a reference time line by setting $tm_1 = 0$.

3.4 Problem Solution

The ML estimate for the node coordinates of the microphones and loudspeakers is implicitly defined as the minimum of a non-linear function. The solution is the same as a nonlinear weighted least squares problem. The Levenberg-Marquardt method is a popular method for solving non-linear least squares problems. For more details on nonlinear minimization refer to [3]. Least squares optimization requires that the total number of observations is greater than or equal to the total number of parameters to be estimated. This imposes a minimum number of microphones and speakers required for the position estimation method to work. Assuming $M{=}S{=}K$, Table 2 lists the minimum K required for the algorithm.

Table 2. Minimum value of Microphone Speaker Pairs (K) required for different estimation procedures (D-Dimension)

$K \geq$	$D=2$	$D=3$
TDOA Position Estimation	5	6
TDOA Joint Estimation	6	7

One problem with minimization is that it can often get stuck in a local minima. In order to avoid this we need a good starting guess. We use the technique of metric multidimensional scaling (MDS) [17] to get a closed form approximation for the microphone and speaker positions, which is used as a starting point for the minimization routine. MDS is a popular method in psychology and denotes a set of data-analysis techniques for the analysis of proximity data on a set of stimuli for revealing the hidden structure underlying the data.

Given a set of N GPCs, let X be a $N \times 3$ matrix where each row represents the 3D coordinates of each GPC. Then the $N \times N$ matrix $B = XX^T$ is called the dot product matrix. By definition, B is a symmetric positive definite matrix, so the rank of B (i.e. the number of positive eigen values) is equal to the dimension of the datapoints, i.e. 3 in this case. Also based on the rank of B we can find whether the GPCs are on a plane (2D) or distributed in 3D. Starting with a matrix B (possibly corrupted by noise), it is possible to factor it to get the matrix of coordinates X. One method to factor B is to use singular value decomposition (SVD) [12], i.e., $B = U\Sigma U^T$ where Σ is a $N \times N$ diagonal matrix of singular values. The diagonal elements are arranged as $s_1 \geq s_2 \geq s_r > s_{r+1} = ... = s_N = 0$, where r is the rank of the matrix B. The columns of U are the corresponding singular vectors. We can write $X^{'} = U\Sigma^{1/2}$. From $X^{'}$ we can take the first three columns to get X. If the elements of B are exact (i.e., they are not corrupted by noise), then all the other columns are zero. It can be shown that SVD factorization minimizes the matrix norm $\| B - XX^T \|$.

In practice we can estimate the distance matrix D where the ij^{th} element is the Euclidean distance between the ith and the jth GPC. We have to convert this distance matrix D into a dot product matrix B. In order to form the dot product matrix we need to choose some point as the origin of our coordinate system. Any point can be selected as the origin, but Togerson [17] recommends the centroid of all the points. If the distances have random errors then choosing the centroid as the origin will minimize the errors as they tend to cancel each other. We obtain the dot product matrix B using the cosine law which relates the distance between two vectors to their lengths and the cosine of the angle between them. Refer to [13] for a detailed derivation of how to convert the distance matrix to the scalar product matrix.

In the case of M microphones and S speakers we cannot use MDS directly because we cannot measure all the pairwise distances. We can measure the distance between each speaker and all the microphones. However, we cannot measure the distance between two microphones or two speakers. In order to apply MDS, we cluster microphones and speakers, which are close together. In practice, it is justified by the fact that the microphones and the speakers on the same GPC are close together. Assuming that all GPCs have at least one microphone and one speaker, we can measure the distance between the speakers on one GPC and the microphones on the other and vice versa. Taking the average we get an approximate distance between the two GPCs. The position estimate obtained using MDS has the centroid as the origin and an arbitrary orientation. Therefore, the solution obtained using MDS is translated, rotated and reflected to the reference coordinate system discussed earlier. Fig. 8 shows an example with 10 laptops each having one microphone and one speaker. The actual locations of the sensors and actuators are shown as 'x'. The '*'s are the approximate GPC locations resulting from MDS. As can be seen the MDS result is very close to the true microphone and speaker locations. Each GPC location obtained using MDS is randomly perturbed to be used as a initial guess for the microphones and speakers on that GPC. The 'o's are the results

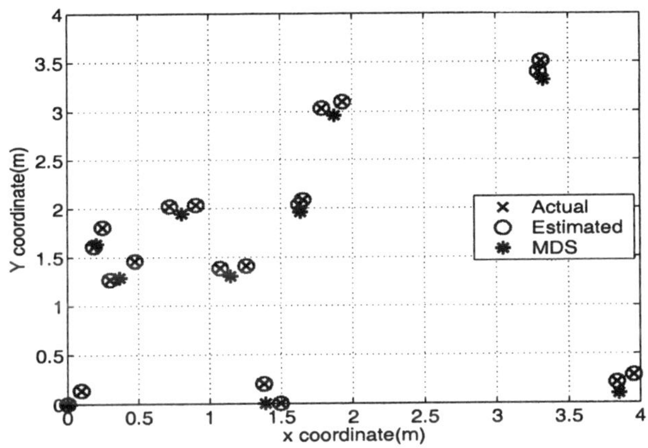

Fig. 8. Results of Multidimensional Scaling for a network consisting of 10 GPCs each having one microphone and one speaker.

from the ML estimation procedure using the perturbed MDS locations as the initial guess. The algorithm can be summarized as follows:

ALGORITHM

Say we have M microphones and S speakers

- **STEP 0**: *Form a Coordinate system by selecting three nodes: The first one as the origin, the second to define the x-axis and the third to form the xy-plane. Also select a fourth node to represent the positive z-axis.*
- **STEP 1**: *Compute the $M \times S$ Time Of Flight (TOF) matrix.*
- **STEP 2**:
  - *Convert the TOF matrix into an approximate distance matrix by appropriately clustering the closest microphones and speakers.*
  - *Get the approximate positions of the clustered entities using metric Multidimensional Scaling.*
  - *Translate, rotate and mirror the coordinates to the coordinate system specified in STEP 0.*
- **STEP 3**:
  - *Slightly perturb the coordinates from STEP 2 to get approximate initial guess for the microphone and speaker coordinates.*
  - *Set an approximate initial guess for the microphone capture start time.*
  - *Minimize the TDOA based error function using the Levenberg-Marquardat method to get the final positions of the microphones and speakers.*

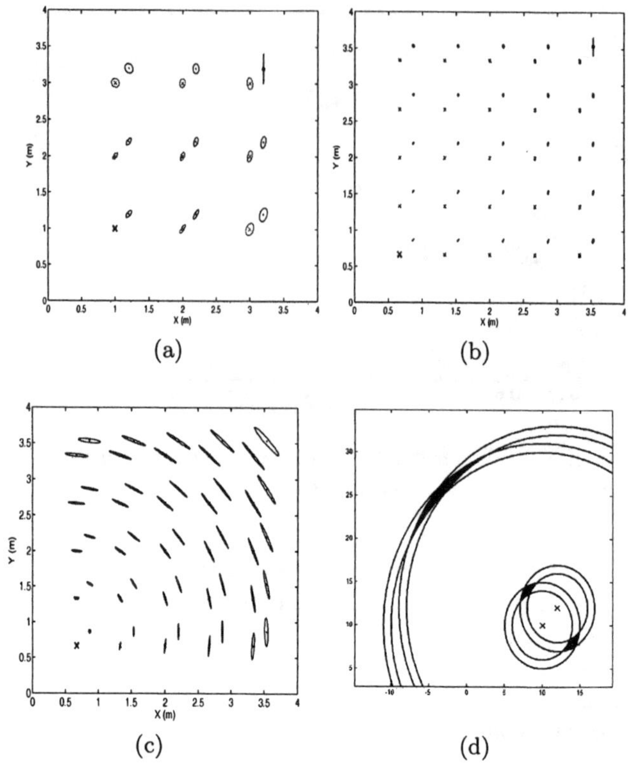

Fig. 9. 95% uncertainty ellipses for a regular 2 dimensional array of (a) 9 speakers and 9 microphones, (b)and (c) 25 speakers and 25 microphones. Noise variance for all cases is $\sigma^2 = 10^{-9}$. The microphones are represented as crosses ($\times$) and the speakers as dots (.). The position of one microphone and the x coordinate of one speaker is assumed to be known (shown in bold). In (c) the known nodes are close to each other and in (a) and (b) they are spread out one at each corner of the grid. (d) schematic to explain the shape of the uncertainty ellipses.

3.5 Analysis

The Cramér-Rao bound (CRB) gives a lower bound on the variance of *any* unbiased estimate [18]. We derived it in [13] for our system leading to the following important observations.

The more microphones and speakers in the network, the smaller the error in estimating their positions as can be seen from Fig. 9(a) and Fig. 9(b) which show the 95% uncertainty ellipses for different number of sensors and actuators. Intuitively this can be explained as follows: Let there be a total of n nodes in the network whose coordinates are unknown. Then we have to estimate a total of $3n$ parameters. The total number of TOF measurements available is however $n^2/4$ (assuming that there are $n/2$ microphones and $n/2$

speakers). So if the number of unknown parameters increases as $O(n)$, the number of available measurements increases as $O(n^2)$. So the linear increase in the number of unknown parameters, is compensated by the quadratic increase in the available measurements.

In our formulation we assumed that we know the positions of a certain number of nodes, i.e., we fix three of the nodes to lie in the x-y plane. The CRB depends on which of the sensor nodes are assumed to have known positions. In Fig. 9(c) the two known nodes are at one corner of the grid. It can be seen that the uncertainty ellipse becomes wider as you move away form the known nodes. The uncertainty in the direction tangential to the line joining the sensor node and the center of the known nodes is much larger than that along the line. The reason for this can be explained for a simple case where we know the locations of two speakers (see Fig. 9(d)). A circular band centered at each speaker represents the uncertainty in the distance estimation. The intersection of the two bands corresponding to the two speakers gives the uncertainty region for the position of the sensor. For nodes far away from the two speakers the region widens because of the decrease in the curvature. It is beneficial if the known nodes are on the edges of the network and as faraway from each other as possible. In Fig. 9(b) the known sensor nodes are on the edges of the network. As can be seen there is a substantial reduction in the dimensions of the uncertainty ellipses. In order to minimize the error due to Gaussian noise we should choose the three reference nodes (in 3D) as far as possible.

3.6 Experimental Details and Results

We implemented a prototype system consisting of 6 microphones and 6 speakers. The real-time setup has been tested in a synchronized as well as a distributed setup using laptops. The ground truth was measured manually to validate the results from the position calibration methods.

A linear chirp signal was used to measure the TOF. A linear chirp signal is a short pulse in which the frequency of the signal varies linearly between two preset frequencies. In our system, we used the chirp signal of 512 samples at 44.1kHz (11.61 ms) as our calibration signal. The instantaneous frequency varied linearly from 5 kHz to 8 kHz. The initial and the final frequency was chosen to lie in the common pass band of the microphone and the speaker frequency response. The chirp signal sent by the speaker is convolved with the room impulse response resulting in the spreading of the chirp signal.

One of the problems in accurately estimating the TOF is due to the multipath propagation caused by room reflections. The time-delay may be found by locating the peak in the cross-correlation of the signals received over the two microphones. However this method is not robust to noise and reverberations. Knapp and Carter [5] developed the Generalized Cross Correlation (GCC) method. In this method, the delay estimate is the time lag which maximizes the cross-correlation between filtered versions of the received signals [5]. The cross-correlation of the filtered versions of the signals is called as the Generalized Cross Correlation (GCC) function. The GCC function $R_{x_1x_2}(\tau)$ is computed as [5] $R_{x_1x_2}(\tau) = \int_{-\infty}^{\infty} W(\omega)X_1(\omega)X_2^*(\omega)e^{j\omega\tau}d\omega$ where $X_1(\omega)$, $X_2(\omega)$ are the Fourier transforms of the microphone signals $x_1(t)$, $x_2(t)$, respectively and $W(\omega)$ is the weighting function. The two most commonly used weighting functions are the ML and the PHAT weighting. The ML weighting function performs well for low room reverberation. As the room reverberation increases this method shows severe performance degradations. Since the spectral characteristics of the received signal are modified by the multipath propagation in a room, the GCC function is made more robust by deemphasizing the frequency dependent weighting. The Phase Transform is one extreme where the magnitude spectrum is flattened. The PHAT weighting is given by $W_{PHAT}(\omega) = 1/|X_1(\omega)X_2^*(\omega)|$. By flattening out the magnitude spectrum the resulting peak in the GCC function corresponds to the dominant delay. However, the disadvantage of the PHAT weighting is that it places equal emphases on both the low and high SNR regions, and hence it works well only when the noise level is low.

In practice, the sensors' and actuators' three dimensional locations could be estimated with an average bias of 0.08 cm and average standard deviation of 3 cm (results averaged over 100 trials). Our algorithm assumed that the sampling rate is known for each laptop and the clock does not drift. Our initial real time setup integrates the distributed synchronization scheme using ML sequence as proposed in [8] to resample and align the different audio streams. It has now been converted to use the synchronization scheme presented in Section 2. As regards to CPU utilization the TOA estimation consumes negligible resources. If we use a good initial guess via the Multidimensional Scaling technique then the minimization routine converges within 8 to 10 iterations.

4 Conclusion and Outlook

We presented our novel algorithms for self-synchronization of distributed AV-sensor networks in time (i.e., synchronized I/O) with a precision of the order of μs and for self-localization in space (i.e., 3D spatial coordinates) with a precision of the order of several centimeters. These algorithms when implemented in real-life systems can provide a completely new platform for future exciting research in areas ranging from manufacturing to communications, entertainment (especially games), and many more.

Researchers interested in using the common time and space infrastructure are encouraged to contact the authors for a research prototype of the system implemented for laptops with Intel® Centrino™ Mobile Technology.

References

1. Elson, J., Girod, L., Estrin, D. (2000) Fine-grained network time synchronization using reference broadcasts. 5th Symposium on OS Design and Implementation.
2. Fancourt, C., Parra, L. (2001) The coherence function in blind source separation of convolutive mixtures of non-stationary signals. Proc IEEE Workshop on Neural Networks for Signal Processing, 303–312.
3. Gill, P., Murray, W., Wright, M. (1981) Practical Optimization.
4. Girod, L., Bychkovskiy, V., Elson, J., Estrin, D. (2002) Locating tiny sensors in time and space: A case study. Proc. International Conference on Computer Design.
5. Knapp, C., Carter, G. (1976) The generalized correlation method for estimation of time delay. IEEE Trans. Acoust., Speech, Signal Processing, **ASSP-24(4)**, 320–327.
6. Lamport, L., Melliar-Smith, P. (1985) Synchronizing clocks in the presence of faults. JACM, **32(1)**, 52–78.
7. Lienhart, R., Kozintsev, I., Wehr, S. (2003) Universal synchronization scheme for distributed audio-video capture on heterogeneous computing platforms. Proc 11th ACM Conf on Multimedia, 263–266.
8. Lienhart, R., Kozintsev, I., Wehr, S., Yeung, M. (2003) On the importance of exact synchronization for distributed audio processing. Proc. IEEE Int. Conf. Acoust., Speech, Signal Processing.
9. Mills, D. (1991) Internet time synchronization: the network time protocol. IEEE Tran Comm, **39(10)**, 1482–1493.
10. Mock, M., Frings, R., Nett, E., Trikaliotis, S. (2000) Clock synchronization for wireless local area networks. IEEE 12th Euromicro Conference on Real-Time Systems (Euromicro RTS 2000), 183–189.
11. Moses, R., Krishnamurthy, D., Patterson, R. (2003) A self-localization method for wireless sensor networks. Eurasip Journal on Applied Signal Processing Special Issue on Sensor Networks, **2003(4)**, 348–358.
12. Press, H., Teukolsky, S., Vettring, W., Flannery, B. (1995) Numerical Recipes in C The Art of Scientific Computing. Cambridge University Press, 2 edition.
13. Raykar, V., Kozintsev, I., Lienhart, R. (2003) Self localization of acoustic sensors and actuators on distributed platforms. International Workshop on Multimedia Technologies in E-Learning and Collaboration (WOMTEC).
14. Rousseeuw, P. (1984) Least median-of-squares regression. JACM, **79**, 871–880.
15. Sachar, J., Silverman, H., Patterson, W. (2002) Position calibration of large-aperture microphone arrays. Proc. IEEE Int. Conf. Acoust., Speech, Signal Processing, 1797–1800.
16. Savvides, A., Han, C., Srivastava, M. (2001) Dynamic fine-grained localization in ad-hoc wireless sensor networks. Proc. International Conference on Mobile Computing and Networking.
17. Torgerson, W. (1952) Multidimensional scaling: I. theory and method. Psychometrika, **17**, 401–419.
18. Van Trees, H. (2001) Detection, Estimation, and Modulation Theory, **1**. Wiley-Interscience.
19. Weiss, A., Friedlander, B. (1989) Array shape calibration using sources in unknown locations-a maxilmum-likelihood approach. IEEE Trans. Acoust., Speech, Signal Processing, **37(12)**, 1958–1966.

Printing: Strauss GmbH, Mörlenbach
Binding: Schäffer, Grünstadt